No. 487
$6.95

64 HOBBY PROJECTS FOR HOME & CAR

By Robert M. Brown
and
Mark Olsen

TAB BOOKS
BLUE RIDGE SUMMIT, PA. 17214

FIRST EDITION

FIRST PRINTING — OCTOBER 1969

Copyright © 1969 by TAB BOOKS

Printed in the United States
of America

Reproduction or publication of the content in any manner, without express permission of the publisher, is prohibited. No liability is assumed with respect to the use of the information herein.

Library of Congress Card Number: 69-14557

Contents

Project	Title	Page

PROJECTS FOR THE HOME

1 Light-Sensitive Garage Light Control — 31

2 Wireless TV Commercial Extinguisher — 35

3 Fire Alarm With Low-Voltage Sensors — 37

4 High/Low Lamp Intensity Control — 41

5 Simple Lawn Sprinkler Control — 44

6 Liquid Level Control — 48

7 Remote Switch For TV, Lamps — 53

8 Low-Voltage Closet Wiring	56
9 High/Low Soldering Iron Control	58
10 Low-Voltage Lap Counter For Slot Cars	61
11 Effective Home Burglar Alarm	64
12 Rain Detector & Alarm	67
13 Shooting Gallery Game	69
14 Audible Continuity Tester	72
15 Automatic On/Off Precision Drill Switch	75
16 Remote On/Off Motor Controller	77
17 Electric Candle That Lights With A Match	80
18 Add-A-Switch Using Exposed Wiring	83
19 No-Load Safety Protector for DC Supplies	85
20 Magic Wand Lamp Switch	87

21 "Ring" That Opens An Electric
Door Lock   90

22 Splash Alarm For Swimming Pool   93

23 Press-to-Talk Tape Recorder Switch   96

24 Low-Voltage Thermostat for
Attic Fan   99

25 Mat Switch for 117v AC   102

26 Coin-Actuated Electric Switch   105

27 Automatic Switch That Turns On
Second Lamp   108

28 Foolproof Smoke Alarm   110

PROJECTS FOR THE CAR

29 Electronic Tach For $5   115

30 Relay-Type Headlight Alarm   117

31 Wake-Up Alarm For Sleepy Drivers   119

32 Fox Hunt Transmitter Sniffer   121

33 Idle Speed Calibrator & Tach 123

34 Six-Transistor Car Reverb 125

35 Universal Safety Flasher 128

36 Vibrator "Substitutor" 130

37 Junk Car Radio & Tape Player Retriever 131

38 VOM-To-Dwell Meter Converter 133

39 Regulator Interference Killer 135

40 The "Surpriser" Theft Device 136

41 Transistorized Mobile Voice Control 137

42 Inexpensive 117-Volt AC Inverter 140

43 Auto Ice Alarm 143

44 250-Volt Mobile Power Supply 145

45 Subminiature Tachometer 148

46 Simple Dwell Meter 151

47 89¢ Direction-Finding Antenna — 153

48 One Transistor Auto Light Minder — 155

49 Powerful In-Car PA System — 157

50 Inexpensive Battery Charger — 159

51 Soldering Gun For Your Car — 162

52 Cheap Light Alarm — 164

53 Automatic Garage Light — 166

54 Economy Door/Hood/Trunk Alarm System — 168

55 Mobile RF Power Meter & Dummy Load — 169

56 Foolproof CB Direction Finder — 172

57 Vibrator Rejuvenator — 173

58 "S" Meter For Mobile Receivers — 175

59 Delay-Action Foilproof Burglar Alarm — 177

60 Generator "Hash" Eliminator 179

61 Handy Timing Light 180

62 Accurate Miles-Per-Gallon Meter 183

63 Double-Purpose Siren 185

64 Turn Signal Beeper 187

Introduction

This collection of construction projects is designed for the hobbyist who is looking for ways to add conveniences to his home and car, and at the same time enjoy his avocation. The projects for the home—in addition to being fun to build—are valuable for their practical use. The newly-developed AC isolation relay, used as the basis for all the home projects, is extremely versatile. A single relay can be switched from project to project, providing an extraordinary buy for the money. (More on that later.)

The auto projects cover a wide range of uses, too. In fact, everything required for a do-it-yourself service center can be assembled. For example: a battery charger, dwell meter and tachometer, mobile soldering iron, strobe, and a variety of burglar alarms using the switches and alarm device already in the car.

But, before we get to the actual projects themselves, it seems appropriate to review the use of tools, electronic components, schematics, and construction techniques.

TOOLS

The few tools required for the construction of these projects are, for the most part, the simple, home-handyman type generally found around the basement workshop.

Soldering gun: A dual-heat soldering gun with fast-heating copper tips is preferred. Generally the 145/210-watt units are used, although a higher wattage (to 325 watts) is useful in soldering to grounded terminal strip lugs that dissipate enormous amounts of heat before reaching solder-flow temperatures. (See Fig. 1.)

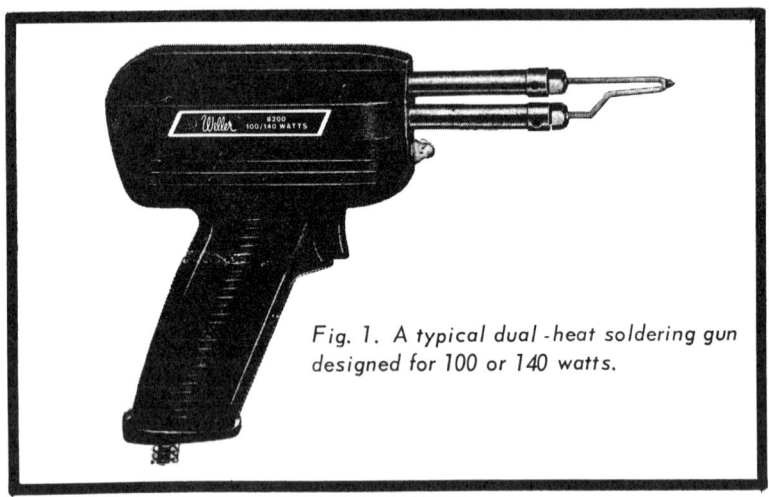

Fig. 1. A typical dual-heat soldering gun designed for 100 or 140 watts.

Solder: Always use rosin-core (radio/TV type) solder. Acid-core types can do more harm than good in an electronics project.

Hookup wire: Insulated (plastic-covered) hookup wire is a must for nearly all connections. For the home projects, where 117v AC is used, heavy-duty single conductor line cord is recommended.

Long-nose pliers: A 5" pair of long-nose pliers is useful both for handling small components and for twisting and looping hookup leads.

Diagonal cutters: Trimming and cutting can be a problem without this aid. Don't depend upon your plier's cutting edge. (See Fig. 2.)

Screwdrivers: Both fairly large and small screwdrivers will be handy for whatever the other tools don't do.

Other tools which are particularly useful in chassis work are:

Drill: A small hand drill, preferably the 1/4" electric type, is invaluable for drilling small holes and even in creating "starts" for larger ones.

Files: A set of three files is recommended for the complete

hobbyist: a rat-tail file, about 1/4" in diameter, a half-round file for smoothing curved edges and a straight-edged file for rough edges.

Other items, not "musts" but extremely useful, include a "nibbling" tool for cutting large holes and a tapered hand reamer for enlarging small round holes. If you intend to go deeper into electronics at a later date, special hole-cutting tools for such things as tube sockets are available; these are called knock-out punches. (See Fig. 3.)

KNOW THE ELECTRONIC COMPONENTS

These are the parts that you will come across most often in your work with the projects in this book:

<u>Resistors</u>: A resistor consumes a portion of the voltage in

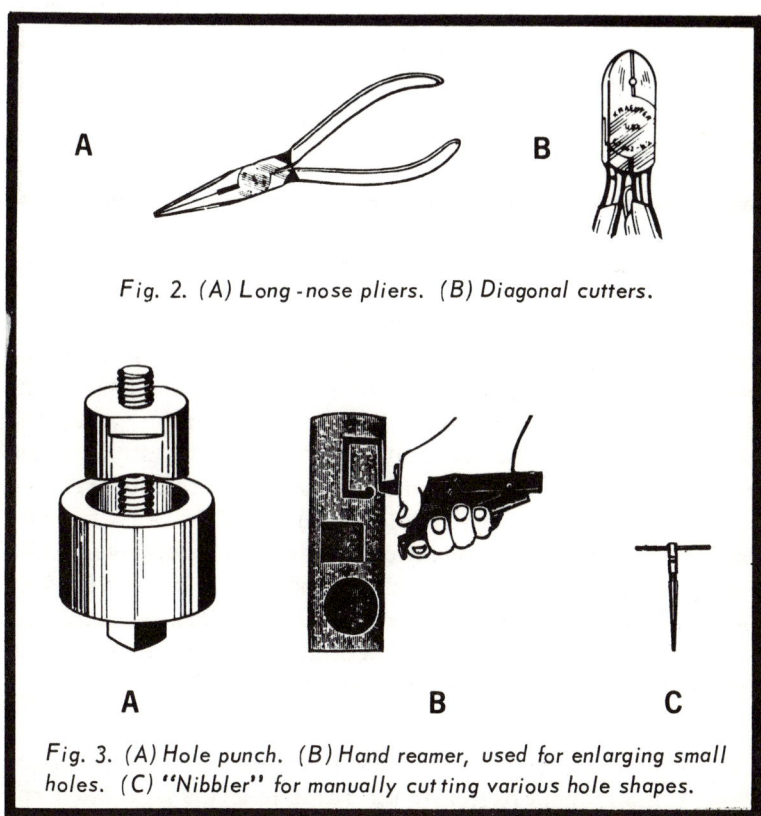

Fig. 2. (A) Long-nose pliers. (B) Diagonal cutters.

Fig. 3. (A) Hole punch. (B) Hand reamer, used for enlarging small holes. (C) "Nibbler" for manually cutting various hole shapes.

11

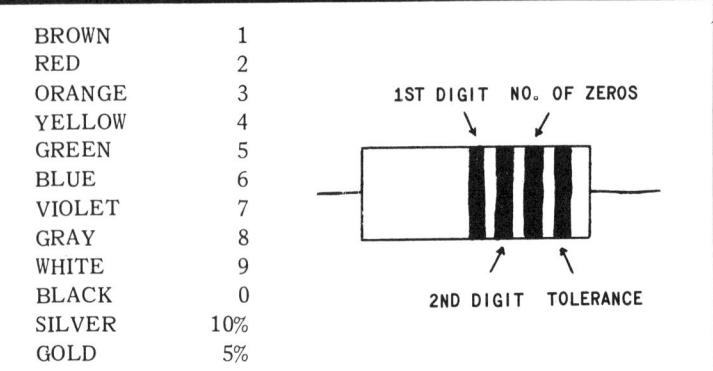

BROWN	1
RED	2
ORANGE	3
YELLOW	4
GREEN	5
BLUE	6
VIOLET	7
GRAY	8
WHITE	9
BLACK	0
SILVER	10%
GOLD	5%

Table 1. Resistor Color Code
A color code is used by most carbon-type resistors to indicate value, each particular color designating a number as shown in the chart. Standard components have three bands to show value. As shown in the diagram, the first band denotes the first digit of the resistance value, and the next band denotes the second digit. The number of zeros following the first two digits is given by the color of the third band. Example: a sequence of brown-black-red means "1" for the first digit, "0" for the second digit and "2" for the number of zeros to be added—giving the resistor a value of 1000 ohms. Similarly, green-blue-black gives 560 ohms. Tolerance is often indicated by a fourth color band. A silver tolerance band means that the actual value of the resistor is within 10% of the coded value; a gold band in the fourth position means 5% tolerance. When there is no fourth band, tolerance is 20%.

a circuit, the amount depending on its rated value in "ohms." In use, it controls voltages and limits current. It is also used for creating potentials with relation to ground. Resistors are generally color-coded, so their values must be interpreted from these bands. (See Table 1.) In the parts lists that accompany each project, resistors are simply presented as: "R2—470K." The "K" stands for thousands, so the rated value is 470,000 ohms. Color bands would then be yellow-violet-yellow. Last band (tolerance) could be silver, indicating ±10% possible variation, which is acceptable. If R2 were listed as 4.7 "meg," the value would be 4,700,000; "meg" means million. Color code? Yellow-violet-blue-silver. Important: Unless specified otherwise in the parts list, resistors are always 1/2-watt types. (See Fig. 4.)

Variable resistors: A variable resistor, or potentiometer ("pot"), performs basically the same function as a fixed resistor, except that its rated value is changeable downward, allowing you to vary the amount of voltage that is dissipated, and, consequently, the "size" of the signal or "loudness" of it (volume control) if it is transformed into sound. (See Fig. 5.)

Capacitors: Capacitors are capable of storing energy and are primarily useful in isolating one component from another; for instance, blocking a DC voltage and letting the AC component pass. Several types of capacitors will be encountered in these projects. Perhaps the most common are the standard tubular or bypass capacitors that appear in parts lists as "C1—.22 mfd"; mfd is for microfarad. Most capacitors of this type have the ".22 mfd" clearly printed on them. Smaller values, however, use color codes (see Table 2). Example ? "C1—110 pfd": This indicates a value of 110 picofarads (which in earlier days of electronics was "micro-microfarads"). Most all variable capacitors are "pfd" types. Example: "365 pfd variable" is a standard AM broadcast radio tuning capacitor for selecting the desired station. Trimmer capacitors, actually screwdriver-adjusted versions of low-value variable capacitors, are also always "pfd" type, although their values are seldom as high as 365 pfd. Trimmer example: "0-16 pfd." If any variable capacitor is simply listed as "25 pfd, as opposed to "0-25 pfd," it is safe to assume that the figure given is simply the highest pfd rating (full capacity). The next type of capacitor you'll encounter is

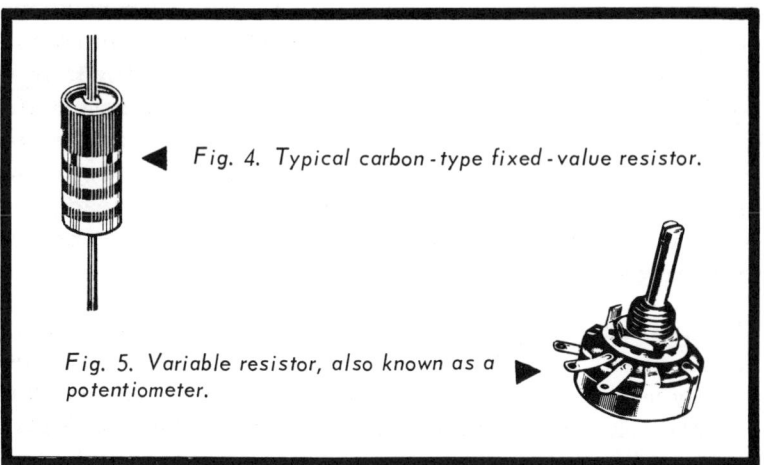

Fig. 4. Typical carbon-type fixed-value resistor.

Fig. 5. Variable resistor, also known as a potentiometer.

Table 44 — Code for Molded Paper Tubulars

Color	1st Significant Figure	2nd Significant Figure	Multiplier	Tolerance %	DC Working Voltage
Black	0	0	1	±20	—
Brown	1	1	10	—	100
Red	2	2	100	—	200
Orange	3	3	1,000	±30	300
Yellow	4	4	10,000	—	400
Green	5	5	—	—	500
Blue	6	6	—	—	600
Violet	7	7	—	—	700
Gray	8	8	—	—	800
White	9	9	—	—	900
Gold	—	—	—	—	1,000
Silver	—	—	—	±10	—

Coding for tubular paper capacitors.

TUBULAR PAPER
1ST, 2ND SIGNIFICANT FIGURES
MULTIPLIER
TOLERANCE
2ND, 1ST SIGNIFICANT VOLTAGE FIGURES

the <u>electrolytic</u> <u>tubular</u>. Electrolytics are shown in parts lists as "C1—16 mfd, 150 WVDC electrolytic. Translation: 16 microfarads (mfd) at 150 working volts DC (direct current). Never substitute a WVDC value less than that specified. <u>Important</u>: Always observe polarity as shown in the schematic diagram by a "+" sign. The same sign appears on the capacitor, as do the other ratings (precluding the necessity for cross-checking color codes). Generally, the "+" lead goes into the circuit while the other lead (minus) goes to chassis ground. (See Fig. 6.)

<u>Diodes</u> <u>and</u> <u>rectifiers</u>: These devices "cut out" one half of an AC signal, converting it to DC current. They are frequently used as "detectors" in radios. The schematic symbol reveals polarity—a crucial item to watch out for. (See Fig. 7.)

<u>Transformers</u>: These generally come in two types: 1) power, and 2) audio frequency (AF). AF transformers are used most frequently to link one amplifier stage with the next or it is used to drive a loudspeaker. On the other hand, a power transformer has either a high or low output voltage and is generally used in the power supply section of the unit. Follow parts list recommendations and you'll have no trouble with these components. (See Fig. 8.)

<u>Transistors</u> <u>and</u> <u>tubes</u>: These now familiar components are generally used to amplify minute signals. Some types—the power transistor, for example—is capable of amplifying much larger signals. (See Fig. 9.)

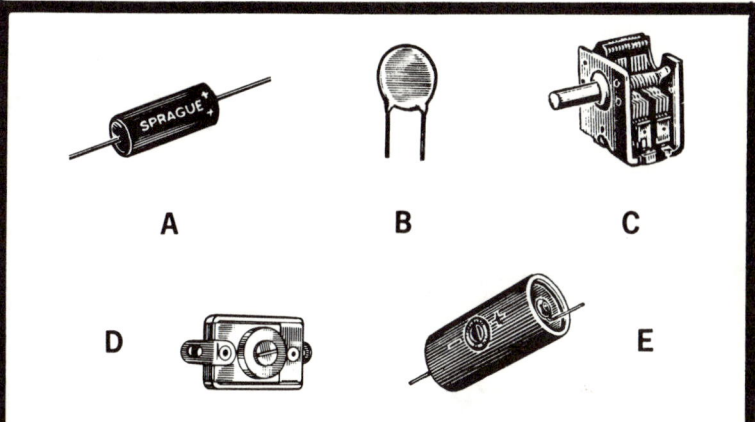

Fig. 6. (A) Standard tubular-type capacitor. (B) A disc ceramic, generally used for bypass applications. (C) A typical variable capacitor, used for tuning purposes. (D) A trimmer capacitor, an infrequently adjusted miniature variable. (E) Electrolytic capacitor; watch the polarity on these.

Fig. 7. One of the many types of diodes. Shown above is a typical silicon power rectifier generally found in the power supply.

Fig. 8. Typical HV (high-voltage) power transformer.

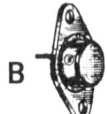

Fig. 9. (A) Standard NPN transistor. (B) Power transistor (for power supply applications). (C) Nine-pin miniature receiving-type tube.

Fig. 10. (A) Magnetic (high-impedance) headset. (B) Crystal (low-impedance) earphone.

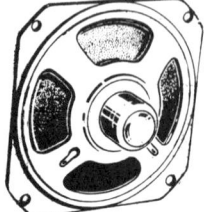

Fig. 11. Inexpensive PM-type loudspeaker.

16

Magnetic and crystal earphones: Magnetic-type earphones are really small speakers operating on low-voltage signals which, of course, convert electrical energy into sound. In use, the magnetic headset is electrically a completed circuit, with high resistance. Crystal earphones—generally small and used in transistor applications—do not complete a circuit. Always check the parts list. (See Fig. 10.)

Loudspeakers: Most are capable of greater power output than any earphone—but work in much the same way (a fluctuating magnetic field vibrating a paper cone). They are generally of extremely low resistances. Example: 4 ohms compared with 2,000 ohms for many magnetic earphones. (See Fig. 11.)

Connectors: All projects should be built with the connector types recommended in the parts list. Power connectors (AC plugs and sockets) are specifically designed to safely handle high-voltage, high-current levels. Test jacks and plugs, on the other hand, are generally used for audio (low-level) signal applications.

BOARDS AND CHASSIS

Most circuits, surprisingly enough, can be mounted on almost any type of base, including many materials you have around the house. Plastic boxes, plywood housing, cigar boxes, aluminum boxes and pegboard pieces are just a sampling of the wide variety that is open to the enterprising experimenter.

Aluminum is perhaps the most popular with electronics hobbyists and is especially suitable for many of the projects in this book because of the stability and rugged protection it provides for the circuit. It is much more workable than steel and can often serve as both cabinet and chassis. Also, a circuit constructed on pegboard is easily secured inside a suitable housing. Special clips (binding posts) are available which can be inserted into board holes. Components and leads fasten directly to these termination points. If a non-conducting material is used, remember to connect all "ground" terminations in the schematic diagram together to complete the circuit.

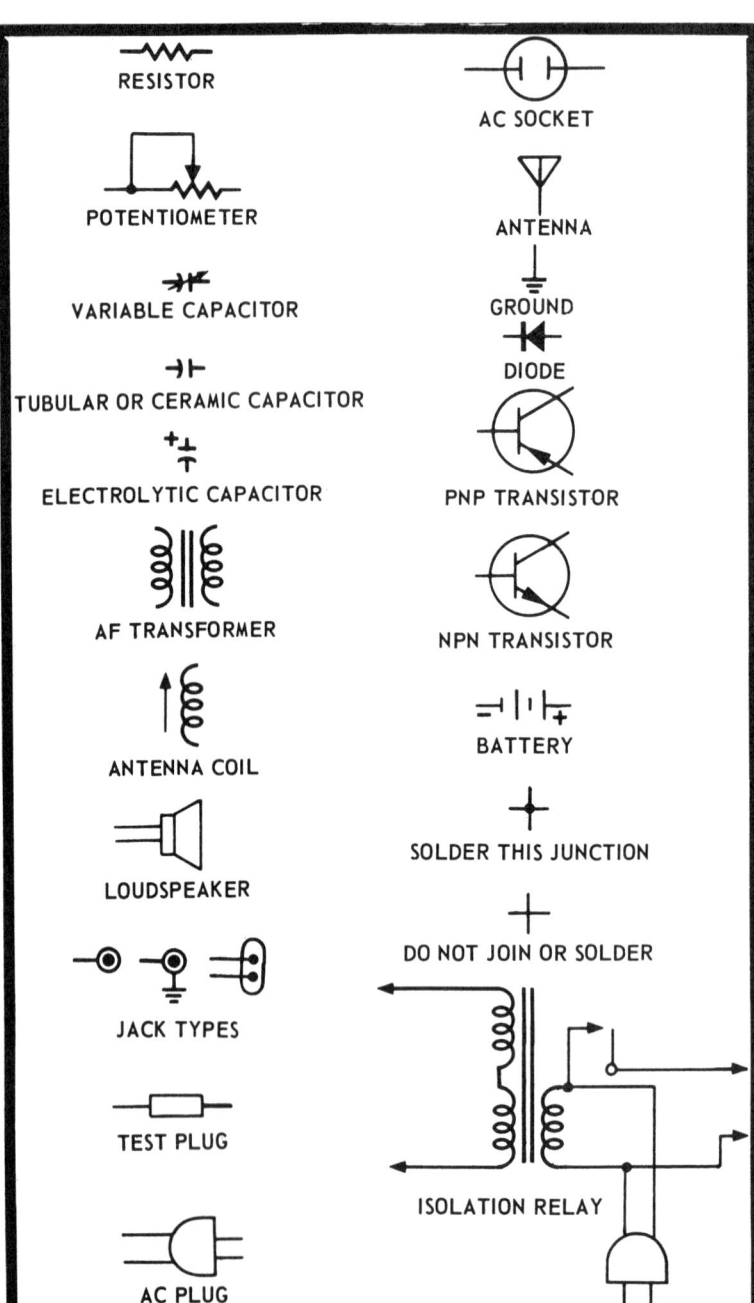

Fig. 12. Electronic components and interconnections are designated by symbols in all wiring diagrams (schematics). For every project a wiring diagram will be provided using these symbols.

While obtaining the desired aluminum sheet size is no problem, hole cutting is a major consideration. This is where some of the tools previously mentioned come in handy. For holes less than 1/4" diameter, an electric drill is recommended. For a larger hole, mark it with a pencil and drill a series of small holes spaced close enough to make the hole easy to "punch" out when it is outlined completely. The "nibbling" tool will cut a variety of shapes in large holes to accommodate meters, switches, etc.

Small round holes are conveniently made with a hand reamer. A small drilled hole can be enlarged up to 1/2" diameter with simply a reamer twisting action. Remove any burrs with a rat-tail file. Chassis holes for accepting hold-down screws are made directly with a drill bit. A supply of 6-32, 1/4" nuts and bolts should be all you'll need.

Lay the parts out on the chassis before making the holes; this brings attention to unsuspected obstructions and significantly adds to the professional appearance of your completed project.

SCHEMATIC SYMBOLS

A schematic diagram, in which each component and connection is designated by a symbol, is provided for each project in the book. The experienced hobbyist who knows the symbols and what they stand for can often tell more about the project by just glancing at the diagram than he can from pages of written "how-to" description. Interpreting schematics is a simple and fast operation; anyone even remotely interested in electronics will find them a great aid in understanding exactly what it is that he's constructing. (See Fig. 12.)

INTERPRETING THE SCHEMATIC

A pictorial diagram can be of assistance in understanding practical circuit construction. Provided here (see Figs. 13 and 14) is an illustration of both a schematic and pictorial diagram of Project 20 in this book. Study the schematic carefully. Notice that the DPDT (double-pole, double-throw) switch has six solder terminals. Filter capacitor C2 and resistor R2, running from ground to one of the switch terminals, are also connected to the centertap of the matching

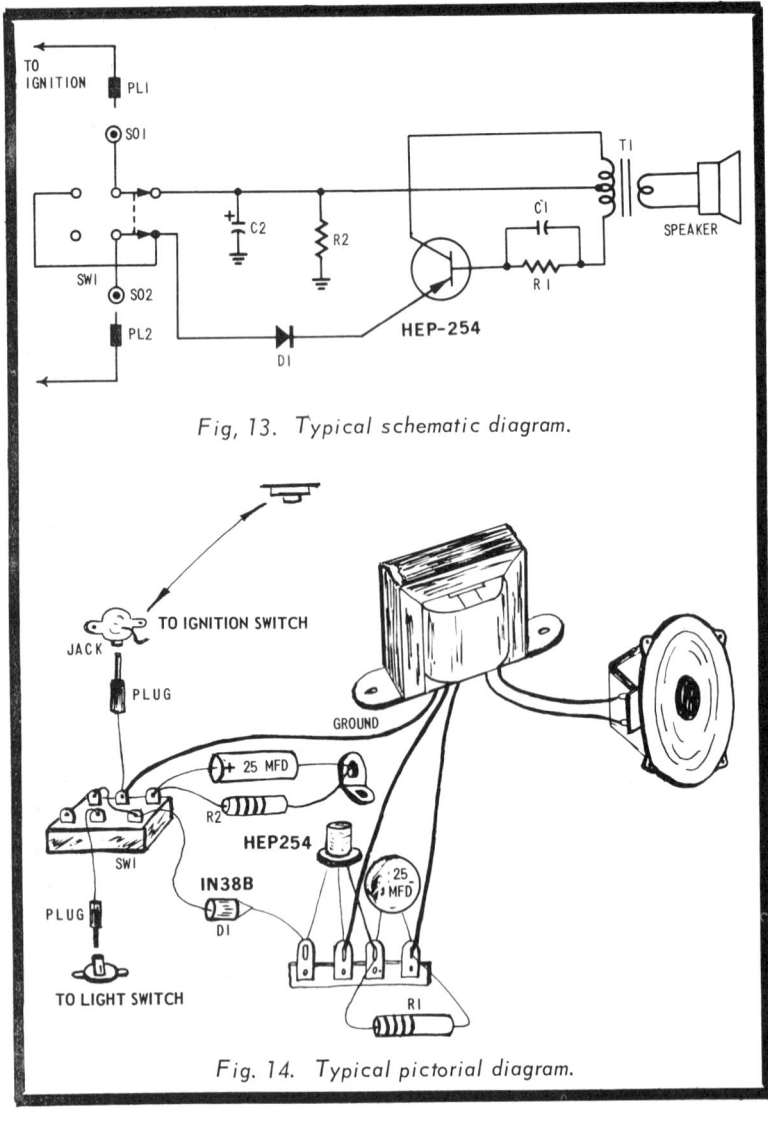

Fig. 13. Typical schematic diagram.

Fig. 14. Typical pictorial diagram.

transformer. A set of test jacks and plugs are connected to the center terminals; these serve as input leads. The transistor is terminated (see the section that follows on "soldering techniques) to both primary transformer input leads; one of these actually consists of a capacitor and resistor.

Look at the pictorial. Notice the circuit relationship to the schematic just discussed. In the schematic only those components vital to circuit operation are shown. The most compact and convenient arrangement for the components is also missing. Rigid points, such as terminal lugs and strips, are needed to make the circuit durable.

Visualizing a pictorial by laying all of the components out on the chassis before hand for any circuit or project that you intend to build will add considerably to the ease of construction. Determine where all of the holes should be for the best arrangement of components and trace them onto onionskin paper or right onto the chassis before you cut them. With a little practice, almost anybody can pick up the knack of expert circuit layout and construction.

SOLDERING TECHNIQUES

Improper soldering will always be the bane of project builders and home repairists. While appearing quite simple, the art of soldering will aid in producing a fine piece of equipment, or be a tremendous help in creating an expensive garbage heap of electrical components. The right amount of heat and the right amount of solder will give you a good joint; only in this way can a circuit operate properly. A little time spent learning correct soldering methods will save time in the long run looking for elusive loose joints. It is more than dropping a blob of solder on two wires.

In addition to soldering gun and solder, you need a small metal cleaning brush, sandpaper, and a soldering aid; this can be a small forked metal rod outfitted with a wooden or plastic handle. Coating the gun's hot tip with solder (called "tinning") helps prevent oxidation, greatly facilitating the soldering operation. Taking this step will prolong the life of the tip. Wipe excess solder off the tip with a wet rag.

Before soldering a joint, the terminal and wires should be freed of accumulated grease and crust with a brush and sandpaper. The joint should be firm enough to remain stable without the solder; wrapping the wires through the terminal at least two times and clamping them firmly into place will assure this stability. The long-nose pliers come in handy here!

When applying solder to the joint, there is an important sequence that should be followed. Heat the joint <u>first</u>. Then apply solder to the joint—<u>not</u> <u>the</u> <u>iron</u>. If this procedure is followed, you will avoid a "cold" joint—a mere blob, where the solder has not actually flowed over the elements and fused them together. A good connection is shiny and ribbed. After trying a few, these are easily identified. Only a little solder is ever required. Practice first.

It should be mentioned here that resistors, transistors, and diodes are extremely sensitive to heat; for this reason care must be taken in soldering their leads. Even a relatively small amount of heat can quickly destroy a transistor. Use a "heat sink"—a metal object attached to the lead between the transistor and the soldering gun to draw off excess heat; an alligator clip works quite well. Sockets on which all soldering can be done before the transistor is installed are convenient and a sure-fire protection against heat damage to the transistor. Beware of filling the socket holes up with solder, however. Another idea, if sockets aren't handy, is to use long-nose pliers as a heat-sink; a rubber band will hold them closed.

GROUND ARRANGEMENTS

All circuits use either a solid ground or a common point, the negative terminal of a battery, for instance. If a wooden chassis is used, there is not too much in the matter. All grounds must be wired together. Keep in mind, however, if a choice has to be made with a circuit where the battery is remote from the unit, that a good external ground, such as a car chassis, makes an ideal return for the usually negative power component.

TROUBLESHOOTING

Troubleshooting is best done during construction. Check each connection at least twice for errors in assembly. If difficulty is experienced after the circuit is complete, have someone else check the connections. An outsider can often find mistakes overlooked by the project builder.

Power polarity—positive to positive, etc.—is extremely critical. Also check for the presence of power! Check the polarity of parts that can be installed in only one direction—electrolytic capacitors, diodes, etc. Switching them can lead to a burn-out and/or incorrect operation of the circuit.

Recheck solder joints. This can be accomplished by holding

each wire and moving it slightly while the circuit is on. Resolder the joint if it is bad.

Short circuits are a frequent cause of trouble. No bare conductors in the circuit should come into contact with each other. Heating up, parts smoking, and fuseblowing are obvious signs of trouble. Correct immediately.

SAFETY

Electricity can be put to exciting hobby uses, but it can also harm—and even kill—if it is handled carelessly. Rules of conduct when working with electrical equipment are quite simple and to the point:

1. Avoid shock. Never touch live component parts or wiring with any part of your body while the circuit is on. Use the common-sense radio-TV technician's practice of putting one hand in the pocket at all times when making screwdriver adjustments. This way, the chance of a shock passing through the body is minimized.

2. Know well the operational circuits of the electronic equipment being built, tested, or used. Know what voltages to expect at different points in the circuit.

3. Avoid "personal grounds." A wooden platform or rubber mat will keep your feet (which are damper than you might think) off the basement's cement floor.

4. Plan all your work carefully.

5. Turn a circuit off and remove the power plug from the wall outlet before replacing parts or tinkering with its internal circuitry.

6. AC powerlines are extremely dangerous—being just the right "frequency" to turn your heart off permanently. Exercise extreme caution with AC equipment containing no isolation transformer.

7. Never trust a mfd-type capacitor. Discharge each one a couple of times if you are working with a disconnected circuit that has been in operation within the last 24 hours. Putting a "bleeder" resistor (high wattage, high ohmic rating) across all filter capacitors is a good idea.

8. Never touch large choke coils until you're certain that power is off and electrolytic capacitors have been discharged. They can deliver a real jolt to the unwary.

9. Quit when tired.

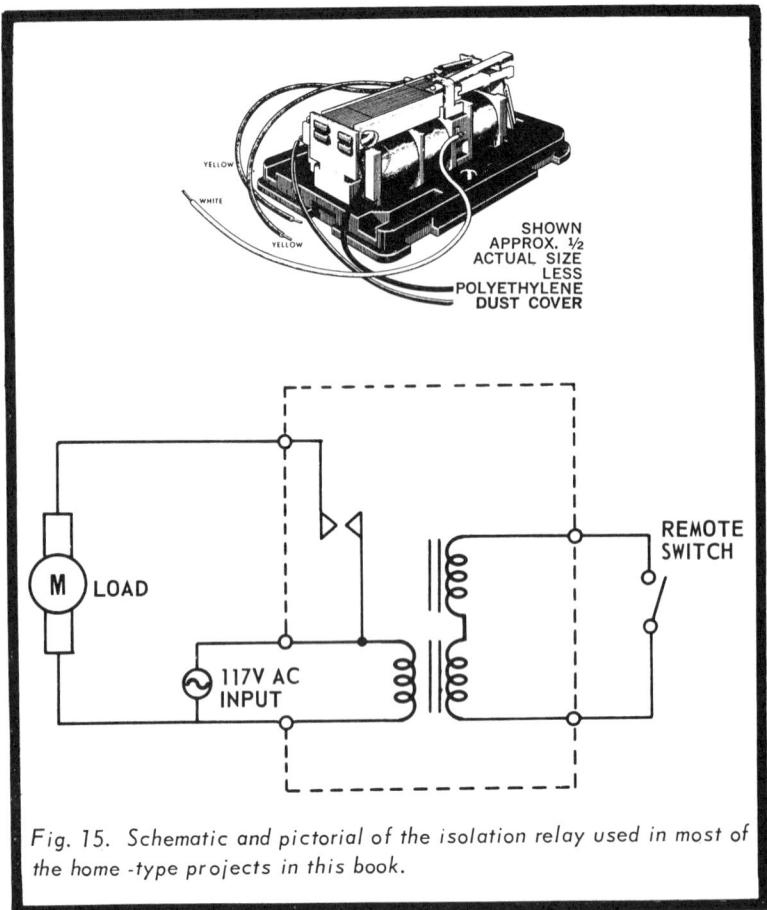

Fig. 15. Schematic and pictorial of the isolation relay used in most of the home-type projects in this book.

PROJECTS FOR THE HOME

An electric lamp that lights with a match; fire, and burglar alarms, TV sound killer, a water level control, and an automatic lawn sprinkler are only a few of the myriad of useful possibilities open to the hobbyist or do-it-yourself'er who owns Alco Electronic Products' Model FR-101 isolation relay. Its low cost ($3.85), extreme versatility and value as a current amplifier, triggering device, and switching device have provided the take-off point for the 28 home projects in this book. Accompanying each project idea is a detailed discussion concerning use, operation, and construction. The experimenter or hobbyist will find that these ideas are really only a beginning and a basis for many other ideas and applications; the only limiting factor is imagination and ingenuity.

The Model FR-101 is a combination step-down transformer and separate relay which share a common core. The result of this combined magnetic device is extremely sensitive operation. 117v AC is applied across the primary, as shown in Fig. 15. The secondary coil's stepped-down voltage is less than 30v AC, and the entire relay draws less than 40 ma. Into the secondary is connected some type of switch, a control device that is either open or represents a short circuit.

The magnetic field induced in the common core is not strong enough to close the relay contacts when no current flows in the secondary winding. When the impedance (resistance, essentially) drops to less than 100 ohms, however, enough current flows to activate the relay and close the contacts. Another interesting property is that an impedance somewhat greater than 100 ohms will pass enough current to force the relay to remain closed, but not quite enough to close it. This idea is employed in several projects in the book, using a capacitor whose reactance (measured in ohms) is very high at 60 Hz.

The rules for safety cited previously cannot be overstressed. While the low-voltage secondary leads can be left exposed usually, it is a "must" to enclose all high-voltage connections. A small aluminum box is perfect in most cases for housing these circuits because of the relay's small size. If a project is made a permanent fixture of the home, be careful to follow local building codes. Where entering an already existing high-voltage line, make the connection in an outlet box, and don't hesitate to use conduit wherever necessary. Where wires pass through holes in metal, rubber grommets will save wear and tear on the wires passing through them.

In moving from one project to the next with the same relay, don't fall victim to careless and hasty put-togethers. High-voltage wiring should always be handled with care and patience to protect the builder and anyone else coming near the circuit. In each circuit are included connectors—test jacks and plugs—to facilitate connecting and disconnecting a unit, and for the polished, expert appearance it gives to the finished circuit.

PROJECTS FOR THE CAR

Everything and anything for the car, from subminiature tachometers to audio reverberation units, are included in Projects 29 through 64. Many projects consist only of a few inexpensive parts; for others you will probably find the makings around the house. It is an opportunity to have some fun and create something useful at the same time.

In addition to the information provided with each car project, the following considerations should prove helpful. First, ascertain the voltage of your car battery and determine whether the positive or negative terminal is grounded. In all lead-acid batteries (the type employed in most cars), each cell is rated at two volts. Count the cells (the caps on the battery) and multiply by two to obtain the voltage. If uncertain, there is often a label on the battery. From one of the battery terminals runs a lead to ground. The sign of this terminal plus or minus is the polarity of ground in the system. It is usually negative. The "hot" connection is the battery's ungrounded terminal.

Power can be drawn directly from the terminal post on the battery, but this is seldom done. The chance of leaving it connected inadvertently after the engine is turned off is too great. The accessory terminal—the longest screw on the back of the ignition switch—is an excellent place to hook in. Removing the key from the ignition cuts the power to the terminal. Also possible is splicing into a lead already connected to the accessory terminal. So little current is drawn generally that there is seldom any danger of overloading the accessory circuit.

Don't be put off guard by the prospect of working with the relatively low DC voltages found in most automobiles. Despite the low voltage, a battery is capable of delivering an extremely large current, which is far more lethal than a high voltage. It is current that kills. While the engine is running, the spark plug cables represent an additional hazard. Your shocked and violent recoil from the high voltage into hot exhaust pipes or running fan is an ever present danger. It's better not to work around the engine when it's running.

In constructing circuits that will be used on the road, be especially careful to secure all components to prevent them from rattling and bumping around inside the housing. Rubber grommets in holes that pass wires is an additional safety precaution that should not be ignored. Also of utmost importance is not mounting the device where it will be an obstruction to the driver of the car.

PROJECTS FOR THE HOME

1

Light-Sensitive Garage Light Control

Invariably, garages are gathering places for junk. Anything that is not "stashed" up in the attic, but which looks like it might be useful someday, finds its way out into the garage where it will sit with all of the space-eating household equipment, such as ladders, hoses, etc. The only problem is that you would probably like to squeeze the car in there, too. Coming home and pulling into the garage after dark can turn into an extremely hazardous and inconvenient situation—not so much in getting into the garage, but in getting out of the car safely once you've parked it. Too many garages are served only by a ceiling light with a switch near the garage door but not near the car door where you really need it.

If your name can be placed on this list, you probably have had many a skinned shin and dusty suit due to your daily crossing of your pitch-black obstacle course. Unseen hoses or garden utensils are a hazard to anyone leaving the car, including your family. Suggested here is not a wall switch near your car door, but something even simpler—a light sensor-activator that will turn that ceiling light on in response to the headlights of your car, or even a flashlight. Remaining on after the initial "trigger" source is turned off, the circuit is self-perpetuating.

As do all of the projects in this book, this one operates very simply, and consists essentially of only a few parts. An ordinary photocell is connected across the secondary winding of the isolation relay (see Fig. 15, Introduction). The actual

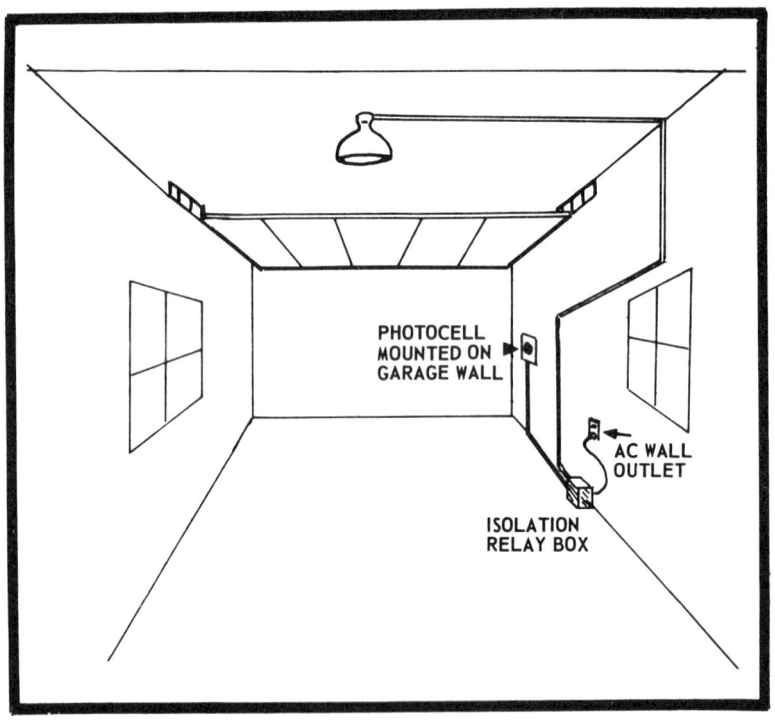

relay, which is in the primary circuit, is connected into one lead of the ceiling light circuit to serve as a switch.

The secondary must be closed to snap the relay shut. The photocell "makes and breaks" this circuit, depending on the amount of light falling on the chemical coated electrode, which emits electrons only when photons of light strike its surface, sending the electrons across the vacuum in the tube to complete the circuit. Since the amount of current through the photocell depends entirely on the intensity of the light striking it, it follows that a strong enough light source will create enough current to trigger the relay and turn the light on. Once the ceiling light is on, it too illuminates the photocell, and with the proper adjustments will keep itself on.

All 117v AC connections should be made inside a small metal box with insulated hookup wire. This is sure to conform to local building safety codes. If your garage is anything like the author's—dank, that is—it's a good idea to use a weatherproofed box to keep out the moisture. Enclosed in the box are

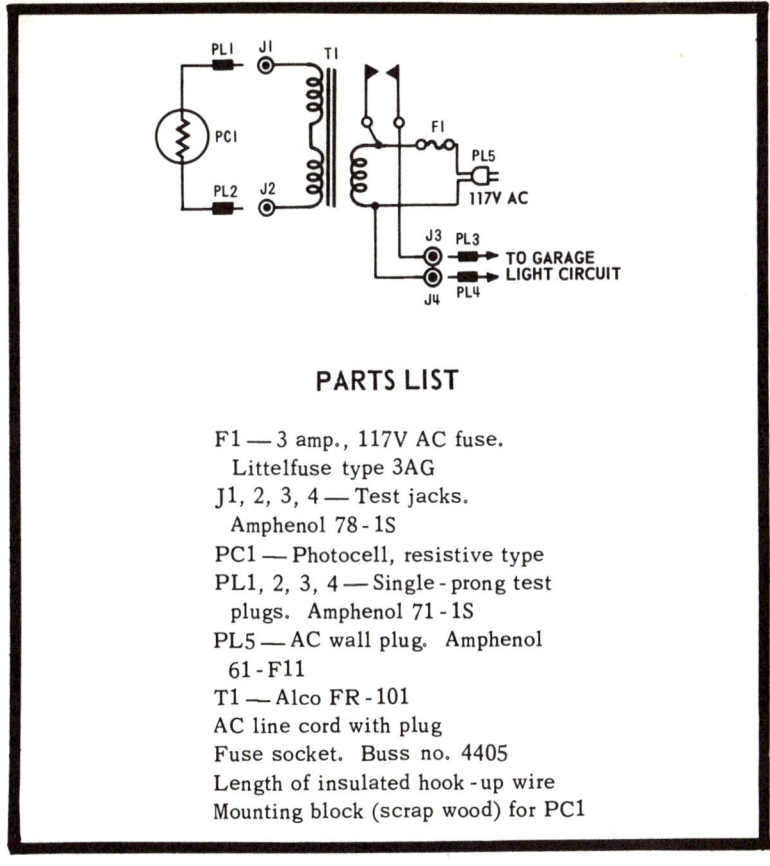

PARTS LIST

F1 — 3 amp., 117V AC fuse.
  Littelfuse type 3AG
J1, 2, 3, 4 — Test jacks.
  Amphenol 78-1S
PC1 — Photocell, resistive type
PL1, 2, 3, 4 — Single-prong test
  plugs. Amphenol 71-1S
PL5 — AC wall plug. Amphenol
  61-F11
T1 — Alco FR-101
AC line cord with plug
Fuse socket. Buss no. 4405
Length of insulated hook-up wire
Mounting block (scrap wood) for PC1

the isolation relay and the fuse, which is provided as an extra safety factor. Insulating all connections with electrical tape, whether they're inside the box or not, is a must. Because the secondary is low-voltage, ordinary bell wire or zipcord will do for the photocell hookup. These leads are safely left exposed and can be just about any length that you need.

There is no foolproof way to break into the ceiling light circuit. The high-voltage lead from the box to the circuit should be as short as possible, and should be heavy-duty AC line cord. Preferably, the connections are made inside one of the outlet boxes. Safest would be mounting the entire switch circuit inside one of these outlet boxes, leaving only the photocell wiring exposed.

Positioning of the isolation relay box and the photocell depends upon the way in which you use the circuit. The back wall of the garage is the place to mount the cell if your headlights will be the "on" signal. If there are windows in your garage, you're apt to run into another problem: light from the windows is going to set off the switch. The solution lies in putting a 2.5K potentiometer in series with the cell to serve as a sensitivity adjustment. Then you can find a happy medium where the light can keep itself on. Placing the cell at the end of a toilet paper tube makes it highly directional. The only limitation is that the main box must be near an AC wall outlet; the photocell can be placed anywhere (see pictorial).

Now you're probably wondering how the thing is turned off. Momentarily turning off the ceiling light will permanently shut it off, since it sustains itself. A normally-closed pushbutton switch in the photocell circuit is perfect, or the already existing wall switch can be flicked from the on to the off position and back to "kill" the light, at the same time resetting it for the next evening. If the relay is wired directly to the light, only the pushbutton will work. Normal wall switch operation is preserved in this way.

A 1.5-mfd capacitor connected across the photocell produces an interesting variation in the operation of the circuit. Once the relay is closed by the photocell, the capacitor will pass enough current to keep it closed. At the very low frequencies involved here, the capacitive reactance is extremely high, creating a barrier to current. The isolation relay does not require as much current to stay closed as it does for the initial triggering action. Consequently, the capacitive effect must be just high enough to keep the relay from closing. This secondary circuit with the photocell, potentiometer, and capacitor combination can be used in a myriad of different ways for adjusting sensitivity and arrangements for keeping the light on.

2

Wireless TV Commercial Extinguisher

TV commercials are one of the most blatant invasions of privacy in our society. You are probably one of those people who is looking for a way to blot out the mind-dulling racket that you have to put up with between program segments. Presented here is a remote sound-control device that will allow you to conveniently cut off the sound during one of these insults to the intelligence.

A small wooden or metal box is ideal for housing the main components, which are the isolation relay, ratchet relay, and fuse. If you place the box on top of your set, as shown in the pictorial, the photocell is best mounted in the front side. To be as inobtrusive as possible, the box must blend well with the TV cabinet. Perfectly all right is hiding the box behind the TV with only the photocell out where you can see it. Since low voltage and current are involved, these leads are safely left exposed. The wall receptacle used for plugging in the TV should be convenient for your device also.

Two relays are used to isolate the speaker leads and voltage from the 117v AC used in the primary of the isolation relay. Speakers, of course, are operated on much lower voltage than this. The ratchet relay connections to the speaker are shown in the pictorial. This relay responds to short pulses of current—one pulse, it opens—another pulse and the relay closes.

Remember, the whole idea of the project is remote control. An essential part of the circuit is a directional light source, brighter than the other lights in the room, used to complete the photocell circuit. The best idea is a light bulb and re-

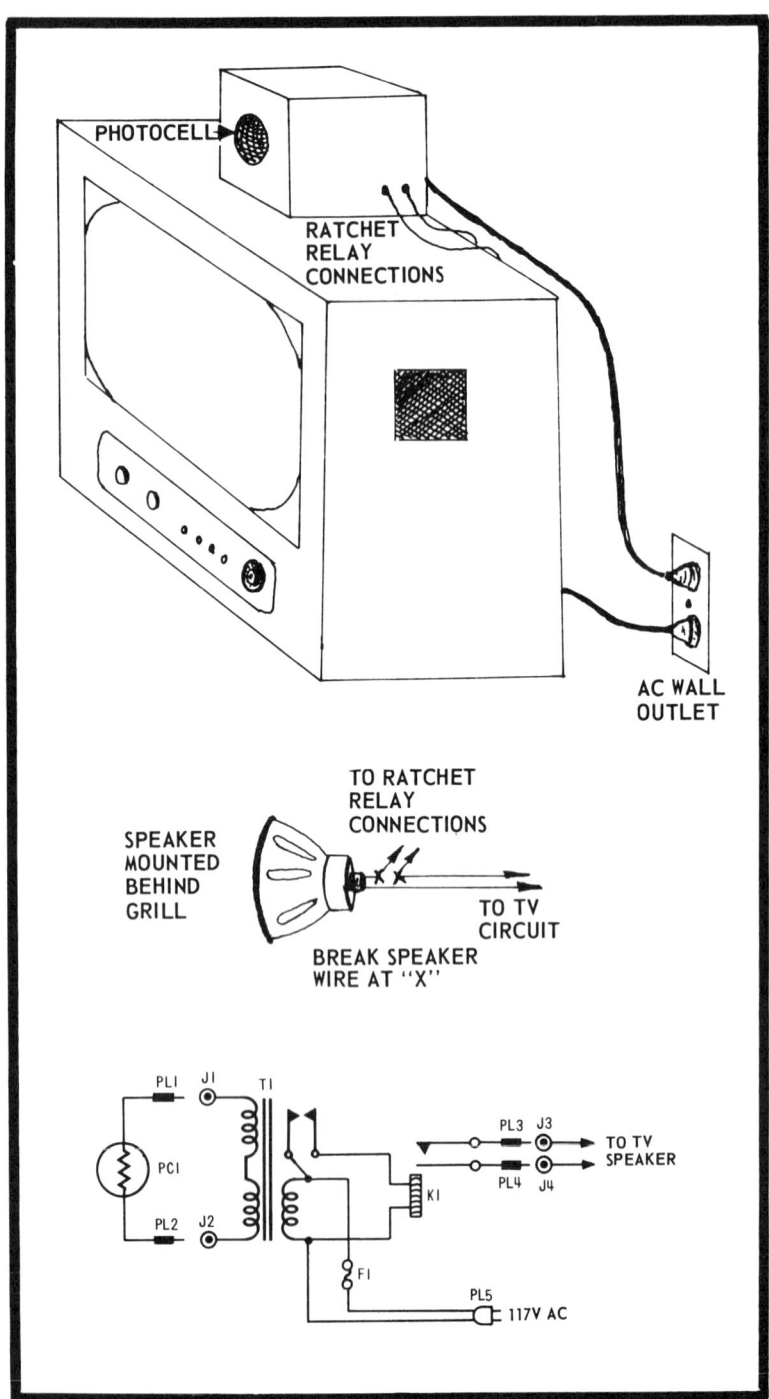

PARTS LIST

F1 — 3 amp, 117V AC fuse. Littelfuse type 3AG
J 1, 2, 3, 4 — Test jacks. Amphenol 78-1S
K1 — Ratchet relay. Alco PC-11A, or Potter & Brumfield type, AP-11A
PC1 — Resistive photocell. Lafayette 19 H 2101
PL1, 2, 3, 4 — Single-prong test plugs. Amphenol 71-1S
PL5 — AC wall plug. Amphenol 61-F11
T1 — Alco FR-101
Length of hook-up wire (insulated)
Fuse holder. Buss no. 4405

flector mounted in a small box that can be aimed once and then used again and again without re-aiming. A simple pushbutton switch in the top is pressed momentarily to turn it on and off. Either a battery or plug-in can serve as power for the light. You will want to secure your source in some way to prevent it from moving.

An alternative light source is a flashlight, flashing it at the photocell before and after a commercial. You will, however, find yourself getting tired of this. The stationary pushbutton suggested above is much easier to use.

Fire Alarm With Low-Voltage Sensors 3

Can you be sure that you don't have a fire in your house right now, putting in dire danger not only you and your possessions, but also your loved ones? Do you smell smoke? Can you be sure? Should you take the chance of not knowing? The answer must unquestionably be "no." Electrical shorts resulting from faulty appliances or wiring are frequently the cause of fires and occur in places where you least expect them. You are probably thinking right now how nice it would be to be able to safeguard just about every corner of the house without too much expense. This project fits the bill to perfection.

37

PARTS LIST

BZ1 — 117V AC buzzer or alarm, hardware-store variety
F1 — 3 amp., 117V AC fuse. Littelfuse type 3AG
J1, 2, 3, 4 — Test jacks. Amphenol 78-1S
PL1, 2, 3, 4 — Single-prong test plugs. Amphenol 71-1S
PL5 — AC wall-plug. Amphenol 61-F11
T1 — Alco FR-101
TS1, 2, 3 — Normally-open thermostatic sensors. GE-X15 thermistors
Fuse holder. Buss no. 4405
Length of AC line cord
Length of insulated hook-up wire
Small metal box

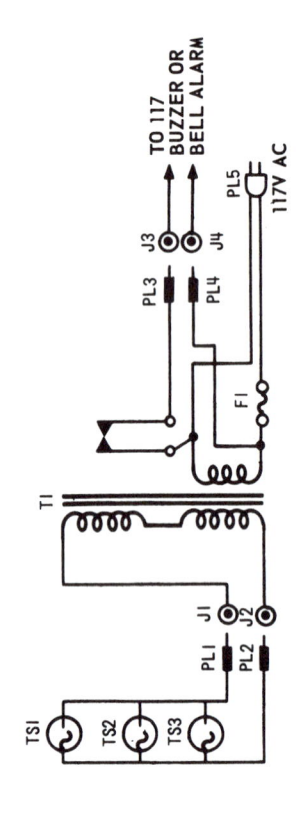

Where there's fire, there's heat, and lots of it. The alarm system described in this project is set off by fire with thermostatic sensors. The other components in the project are the isolation relay (see Introduction) and a 117-volt buzzer or bell alarm. The sensors are normally-open switches that respond quickly to dangerous heat levels. Each one contains a bi-metallic strip that is highly temperature sensitive. One of the metals in the strip has a higher coefficient of linear expansion, bending the strip either to close or to open the connection, depending upon the direction of the temperature change.

The operation of the circuit is very simple. An alarm device is connected across the relay in the primary of the transformer. In the low-voltage secondary are connected the sensors, connected in parallel so that if any one of them "kick off" the secondary will be complete. Then, the relay will snap shut and activate the alarm device. If the sensors were in series, every one of them would have to be closed by the heat for the alarm to go off. This, of course, defeats your purpose of catching a fire before it goes too far. An unlimited number of sensors may be connected in parallel throughout the house for the ultimate in protection.

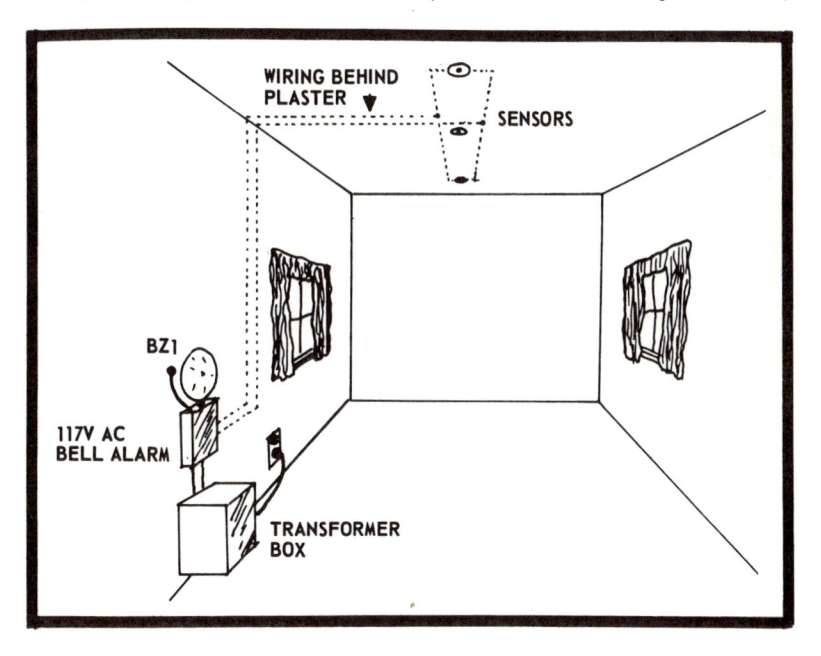

Enclose the heart of the circuit, the isolation relay and fuse, in a small metal box for safety purposes. All 117v AC connections inside the box should be made with insulated hookup wire. And, so as not to mar the appearance of a room, you will want to do the sensor wiring behind the wall and ceiling, with only the sensors exposed. Locations for the sensors are up to you. The ceiling, as opposed to some remote corner of the room, is always a good place because it's out of the way and in a location where it will benefit from the hot air if there is a fire. Two sensors are quite enough for a small room, but in a larger room you may need one or two more. Wiring sensors into more than one room is perfectly all right, too. The only disadvantage to this is the increasing length of the wires as you get more and more sensors in the circuit, and then, too, the alarm will not be localized. Any one of several could be "hot." It shouldn't be too hard to find out, though.

Place the box and the alarm wherever you like, but keep them together to minimize exposed AC wiring. The alarm probably should be someplace where it can be heard no matter where you are in the house. The kitchen or some other central location is fine. Don't make the mistake of muffling the sound by sticking it away in a closet, though.

Fire becomes a greater hazard than usual when your vacationing away from home. If a fire does start, the neighbors might not detect it until your house is on the way to the ground. You can help matters by placing a fire alarm outside. Let your neighbors know what you've done so they can react promptly if ever they should hear it and you're not home.

If you have children, they—more than anyone else—should know how to respond to the alarm. If you aren't home, their knowing the number of the local fire department may save the day. The fire department does not have eyes everywhere. The obligation of doing everything possible to safeguard your family and possessions is your own.

4   High/Low Lamp Intensity Control

Nothing but headaches result from watching TV by the glare of a bright lamp. The irksome glare on the screen is hard on both the nerves and the eyes. If the lamp does not have two or three levels of intensity, allowing you to choose the one that best suits the occasion, what can you do? Turning the lamp off and watching the television in complete darkness is just as bad. Now, with this project you can cut the light intensity in half by the touch of a switch. Late at night, if for some reason you have to turn the light on, a much lower intensity is much easier for your eyes to adjust to. How often do you find yourself stumbling around in the dark rather than subject your eyes to such a drastic change? Never more! Just plug the lamp into the high-low switch circuit and your problems will be over.

The "core" of the project is the silicon rectifier, D1, and the isolation relay. The type of rectifier that you select depends entirely on the wattage of the bulb that you will use. It should have a rating of 1-1/2 amperes for every 100 watts of the bulb. It should be noted here that the switch will not turn the lamp off, but will only vary the intensity. The lamp is turned on and off in the usual manner.

As you are able to see from the schematic, there is always a closed circuit from the 117v AC outlet, through the rectifier, to the lamp. Consequently, whenever the lamp is turned on, it will light. With the switch in the dim position one half of the AC signal is cut out, as the current can flow in only one direction. Greatly reduced intensity in the lamp is the result.

41

PARTS LIST

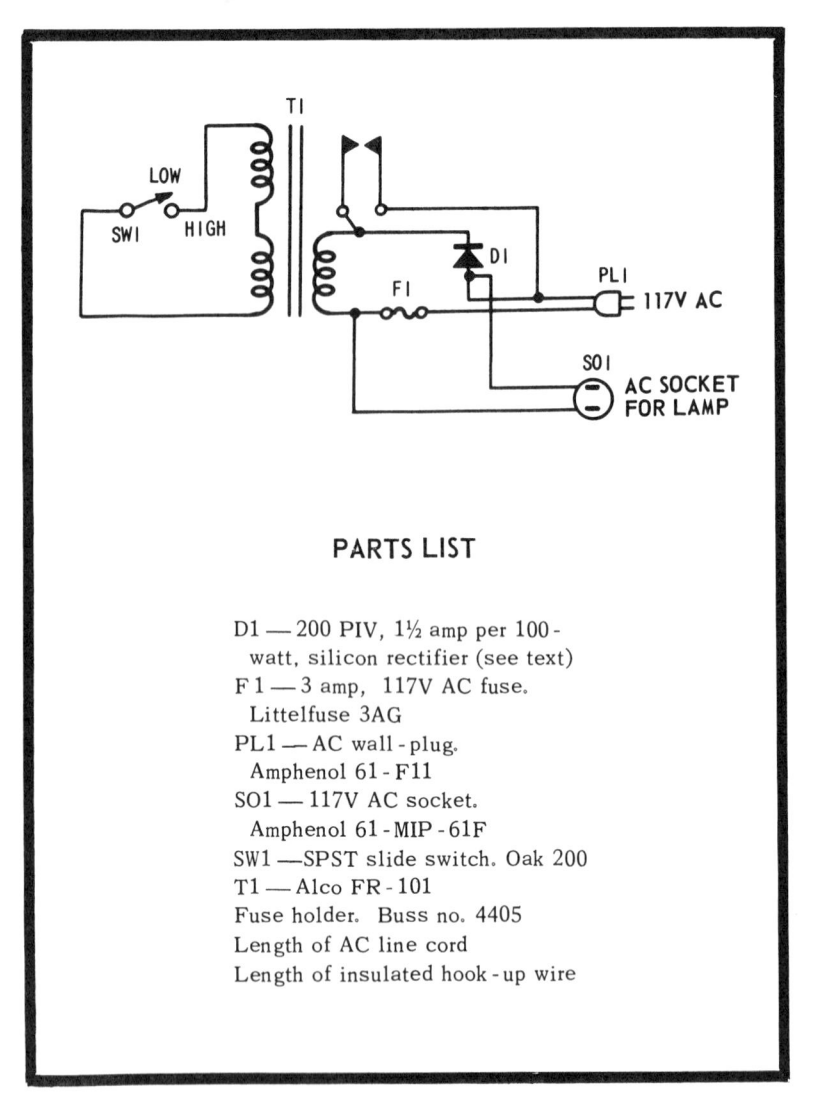

D1 — 200 PIV, 1½ amp per 100-
watt, silicon rectifier (see text)
F 1 — 3 amp, 117V AC fuse.
Littelfuse 3AG
PL1 — AC wall-plug.
Amphenol 61-F11
SO1 — 117V AC socket.
Amphenol 61-MIP-61F
SW1 — SPST slide switch. Oak 200
T1 — Alco FR-101
Fuse holder. Buss no. 4405
Length of AC line cord
Length of insulated hook-up wire

When the relay in the primary circuit is closed, a free path is offered to the current, increasing intensity to normal. The slide switch in the secondary, of course, is the key to closing the relay. If the switch is open, "low" current results, going through the diode; if the switch is closed, "high" intensity follows. Simple as pie.

Construct the circuit in a small metal box. The switch and the female socket for the lamp plug are the only exposed components. All other connections are made inside the box with insulated hookup wire. AC line cord goes to the wall outlet. The voltage in the secondary is very low, making operation of the slide switch absolutely safe. Also, a fuse is included in the circuit for added safety. The circuit can be used as a remote control switch, within the limits of the lamp cord length.

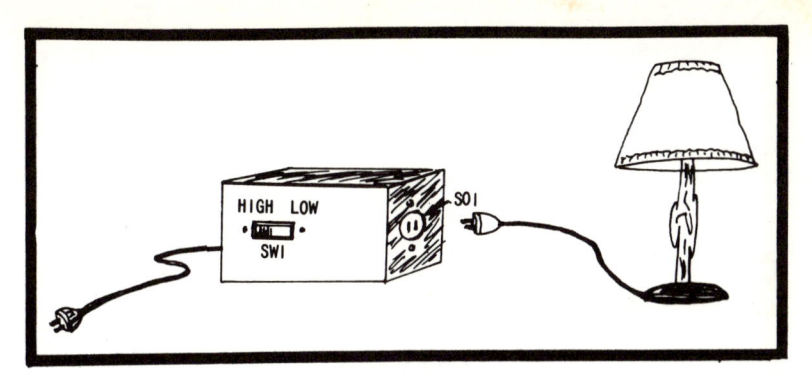

Another idea for using this intensity-lowering technique is with tropical fish tank lights. Anyone who has ever persued the hobby realizes the danger involved in turning a light on over a dark tank. The ideal is going up in intensity in stages. This intensity switch solves the problem. Or maybe you just want to relax or sleep for a while with a dim light on. You can't go wrong with this switch. Appliances other than lamps that operate on 117v AC can be controlled this way; e.g., the speed of a motor. The possibilities go on and on. The rest is left to you.

Since the box will be setting out in the open, you may want to make it a little more attractive. Painting, covering it with paper, or enclosure in a wooden cabinet are but a few of the many creative paths open to you.

5 Simple Lawn Sprinkler Control

It is going to be an extremely hot summer day. You haven't had rain for weeks; everything is withering and turning brown —especially your once beautiful lawn—and the air smells like dust. About this time, you decide that it might be a good idea to bring out the old lawn sprinkler to see what you could do to bring things back to life. So you proceed to set it out in the yard, turn the water on, and then go about your business. Two hours later, glancing outside, you find to your dismay that it looks like you've drowned the lawn rather than saved it. Obviously, the main disadvantage to watering is that the sprinkler cannot be left in one place, but must be moved around or turned off and on periodically to prevent "overwatering." Running in and out to make these adjustments, or watering while you're not home can be a real problem if you don't have an automatic system. This project, which operates on the principle of electrolysis, is the answer to the dilemma.

The circuit consists mainly of the isolation relay, a solenoid valve, and a couple of probes. The solenoid valve, which operates at 117v AC, is an arrangement where a metal core is placed in a wire coil. When connected, the coil pulls on the core and consequently closes the valve. This valve is available at most hardware stores (see parts list). As you can see from the pictorial, the valve is connected between the outside water faucet and the sprinkler.

You probably will want to build your main circuit so that it can be placed outside, near the faucet. This means enclosing it in a completely weatherproofed metal box. The relay and the fuse are the only components that go in the box, where insulated hookup wire should be used to make the connections. Also, all solder joints should be covered with electrical tape, just in case some water does seep into the box. Of course, it is possible to place the box inside the house and run the wires out. It is up to you. The leads to the probes and all other wire that will at any time be exposed to the weather should be outdoor-type insulated wire. Heavy-duty AC line cord is used for the connection to the wall outlet.

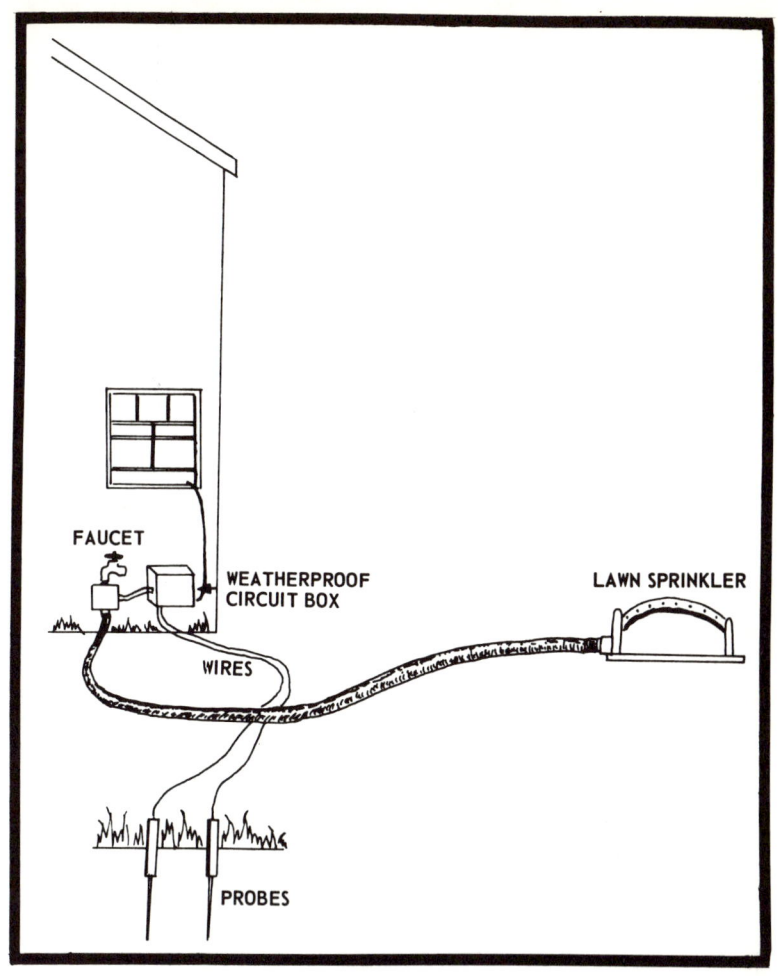

Any metal that won't corrode too much under the moist conditions out in the yard will serve nicely for the probes. The wire should be partially insulated to make the portion of the probe at ground level insensitive to moisture. This way, you

PARTS LIST

F1 — 3 amp 117V AC fuse. Littelfuse 3AG
J1, 2, 3, 4 — Test jacks. Amphenol 78-1S
PL1, 2, 3, 4 — Single-prong test plugs. Amphenol 71-1S
PL5 — AC wall plug. Amphenol 61-F11
T1 — Alco FR-101
VL1 — 117V AC solenoid type valve, hardware-store variety
2 Semi-insulated probes
Fuse holder. Buss no. 4405
Length of outdoor-type AC line cord
Length of outdoor-type insulated wire (no. 10 or 12 copper standard)
Weatherproofed metal box to house circuit (see text)

can adjust somewhat the depth at which you will monitor the moisture in the soil.

The circuit operates very simply. The two probes are connected across the open secondary. When the secondary is closed, the solenoid valve is energized and the water is shut off. The secondary is closed when there is enough moisture in the soil to conduct electricity between the probes. The valve will remain shut as long as there is enough moisture in the soil.

Obviously, the degree of water that you want your system to respond to depends entirely on your positioning of the probes. The deeper the probes are in the earth, the more moisture on top it will take to set them off. The adjustment that you make might depend on the mineral content in your soil, too.

Another idea is to take the probes and place them along with some salt in a plastic container. This can then be placed under the sprinkler. The probes, which are taped to the sides of the container an arbitrary distance from the bottom, will be connected only when the water level in the jar rises to that point. This will serve as a timer for you. If you turn the water on early in the morning and want it to go off before the sun gets too hot and starts burning the grass through the "focusing" drops of water, but you won't be around to shut if off, then this system should help you out.

Sprinkling at night is just about the best time possible, but it is not feasible unless you want to stay up all night and move the sprinkler around. Using this project, the water will stay on only as long as the ground needs water. What more could you ask for, other than having the sprinkler move around by itself. The possible applications for this circuit are far too numerous for elaboration here. The others are left to you. Happy sprinkling!

Liquid Level Control   6

The universal problem of homeowners who have built on low ground, and who have to put up with over-head sewers as a consequence, is water in the basement. If it were not for the sump pump which gets the excess water up to the sewer, you would have a perpetually "floating" basement. But how often, nevertheless, have you awakened in the morning or come home from a vacation to find to your grief and despair that the pump has malfunctioned and your newly purchased rugs and furniture down there are floating around in three feet of water. People have contemplated selling a perfectly good house just because they failed to lick the problem of the annual spring basement flood.

Very often, the cause of the pump malfunction is a burned-out switch. These switches are subject to a tremendous amount of wear and tear because of the pump's being turned on and off every few minutes. Another cause of pump failure also is in the switching system; the rod for the float may get stuck, leaving the pump permanently "off." The same for a switch turned "stiff"; or the float may get stuck, leaving you with the same dilemma; a demonstration of the bouyancy of your basement furniture.

You probably have noticed that all the difficulties thus far mentioned are in the switching mechanism of the sump pump. Obviously the only solution is to find an infallible switch—what we have in this project.

There are two circuit arrangements that may be used. The first circuit consists mainly of a length of coaxial cable modified into a water "probe," the isolation relay, and a potentiometer (see Fig. 1). It will do electrically what previously has been done mechanically.

All high-voltage connections must be made inside a small, "tight" metal box. The transformer, fuse, potentiometer, and AC socket for the pump plug are all housed in this box. The pot, which serves as a sensitivity control, and the socket are exposed (see pictorial 1). Insulated hookup wire and electrical tape for the solder connections will add to the durability of your switch under moist conditions. Don't get the idea that it's all right for it to get wet. Keep the box on an elevated shelf, at least someplace where it will never be in danger of being immersed, assuming continuing operation of the pump.

In this project, the switch that "makes and breaks" the secondary is merely coaxial cable. Like most coax, it has four parts: the inner conductor, a layer of plastic insulation, the shield, and a second insulating cover. As shown in the pictorial, there is a portion of each showing on the end that is to be immersed in water. A part of the shield should be exposed, but should be separated from the center conductor by a small exposed interval of the first layer of insulation.

Meanwhile, inside the box the shield of the coax is connected as one half of the secondary circuit for the isolation relay; the center conductor is the other half. When these two are connected, and the secondary closed, the relay is tripped and the pump brought into action. You place the probe at the depth in the sump where you want the pump to start. A bracket to hold the coax in place inside the sump will have to be made. A small plate of metal with a flexible metal band fastened across the cable is recommended. Two pieces of wire will serve to hang the bracket inside. You probably will want to adjust the depth for different seasons of the year. The secondary is low voltage, so there need be no fear of shock in the moist conditions of a basement.

The second circuit is a little more practical. In the first circuit, the pump goes on whenever the water level rises high enough. This is all fine and dandy as long as the water stays up. But when the pump goes on, the water level falls very fast, turning the pump off immediately, then back on again. This creates another problem with wear and tear on the relay.

Two isolation relays are used in the second circuit idea to

Fig. 1

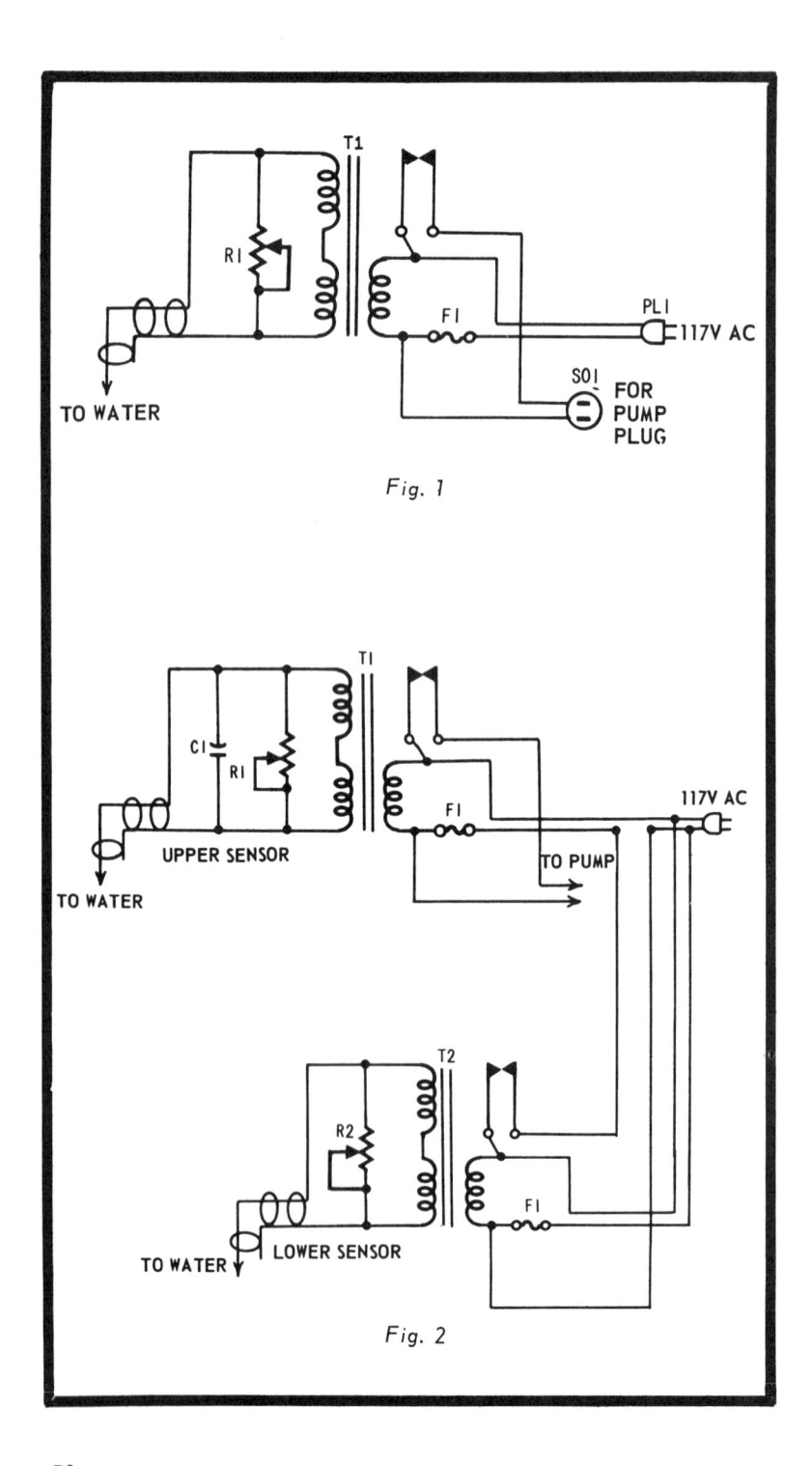

Fig. 2

PARTS LIST

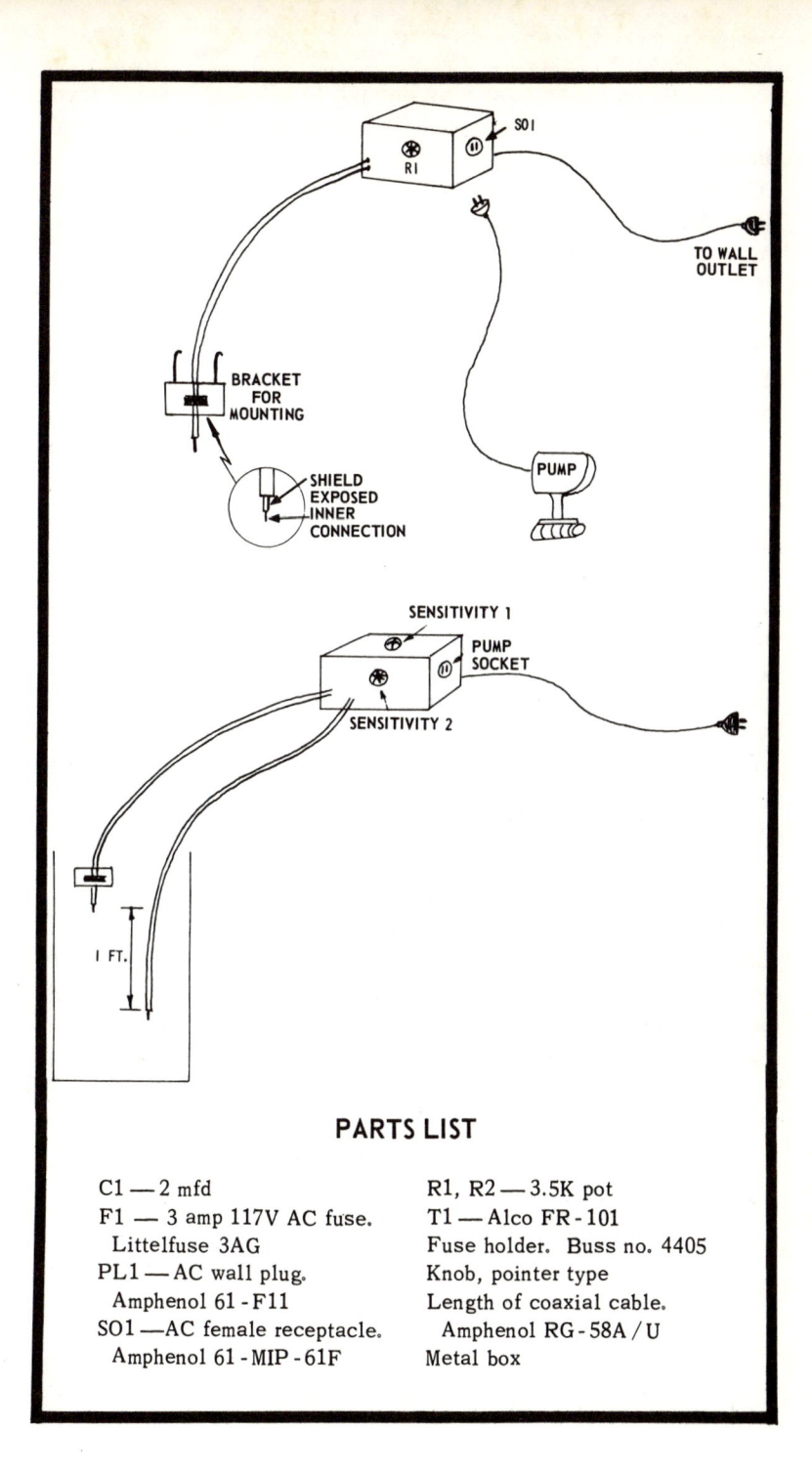

C1 — 2 mfd
F1 — 3 amp 117V AC fuse.
 Littelfuse 3AG
PL1 — AC wall plug.
 Amphenol 61-F11
SO1 — AC female receptacle.
 Amphenol 61-MIP-61F

R1, R2 — 3.5K pot
T1 — Alco FR-101
Fuse holder. Buss no. 4405
Knob, pointer type
Length of coaxial cable.
 Amphenol RG-58A/U
Metal box

keep the pump on at least long enough to allow the water level to drop a foot. This way it will not be going on and off continuously.

Construction is accomplished in the same manner as the first design. Both relays are served by the same plug, and the box has two sensitivity controls, one for each relay (see Fig. 2 and pictorial 2). The probes are made in the same way, also.

The key to success is the 2-mfd capacitor in the secondary of T1. Once the relay is activated by the rising water level, this capacitor keeps it closed even after the water is no longer in contact with the probe. The relay in T2 is connected across the 117v AC line in the T1 primary. The probes are positioned in the sump so that the one from T2 is about a foot below the probe from T1. When the water touches and closes T2, the only thing that happens is a partial completion of the primary circuit of T1. As the water rises as far as T1, the pump goes on and stays on, until the water finally passes the lower sensor, which opens the primary in T1, thus shutting off the pump, which cannot possibly operate beyond this point. A foolproof system.

The two sensitivity controls should be adjusted so that each respective relay is just triggered when immersed in water. Otherwise you're apt to have a direct short with only humid air. Plug the pump in, plug in your switch mechanism, make the necessary adjustments with R1 and R2, and you're finished.

The only moving part in the switch is the relay. Therefore, you have no problem with stuck floats, stuck rods, stiff switches or burned-up switches. The relay life will be much longer than what you are using now and will give continuous operation. You can feel safe for only the minimal cost of the parts for whichever arrangement that you choose to use.

Remote Switch For   7
TV, Lamps

Harry Smith has just finished dinner at home after a grueling day at the office. Mrs. Smith leaves to take the kids to play practice over at the school, so Harry decides to watch a little TV. Only after he turns the TV on and gets settled in a comfortable chair does he realize that he's watching nothing but a glaring spot on the screen caused by the lamp across the room. Not wanting to move, he tries "wishing" it away, but winds up treking over there. It would have been a lot more convenient if he had a switch right at his chair to turn the lamp off.

Settled again in his comfortable chair, he is interrupted by a telephone call which turns out to be a long-lost uncle calling from Outer Slybovia. Harry can't hear him over the racket from the television, so he has to dash back in there to turn it off. A switch to control the television near the telephone would make this frequently occurring situation much less awkward.

These are two of the multitudinous occasions where a low-voltage remote control switch would make the difference. Any appliance or motor that operates on 117v AC can also be controlled in this way. Perhaps you are using a hoist for some heavy job in your basement, and are looking for a safe switch that can be held in the palm of the hand and operated with your thumb. This project is probably just what you're looking for.

Consisting of only a few parts—an SPST slide switch, the isolation relay, an AC socket, and a fuse—this project is constructed in the same manner as many others in this book.

PARTS LIST

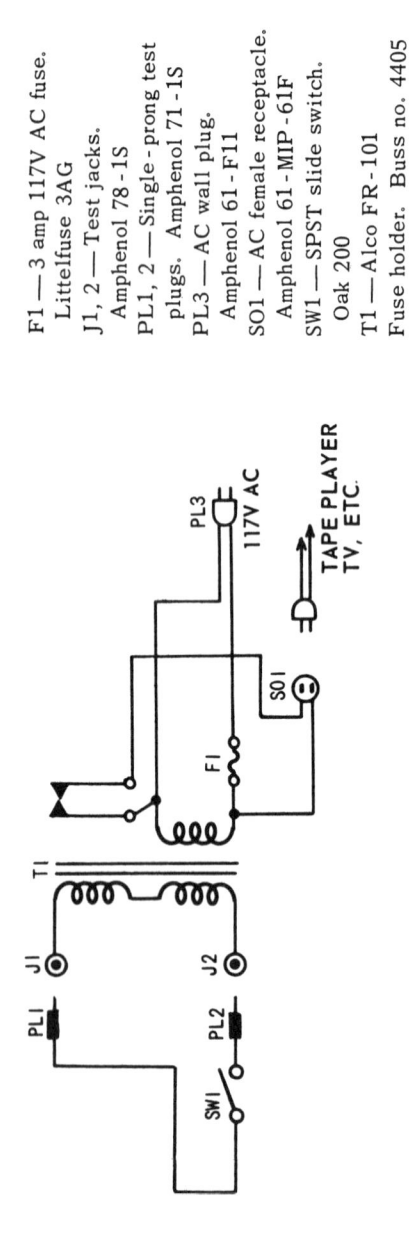

F1 — 3 amp 117V AC fuse.
 Littelfuse 3AG
J1, 2 — Test jacks.
 Amphenol 78-1S
PL1, 2 — Single-prong test
 plugs. Amphenol 71-1S
PL3 — AC wall plug.
 Amphenol 61-F11
SO1 — AC female receptacle.
 Amphenol 61-MIP-61F
SW1 — SPST slide switch.
 Oak 200
T1 — Alco FR-101
Fuse holder. Buss no. 4405
Length of AC line cord
Metal box

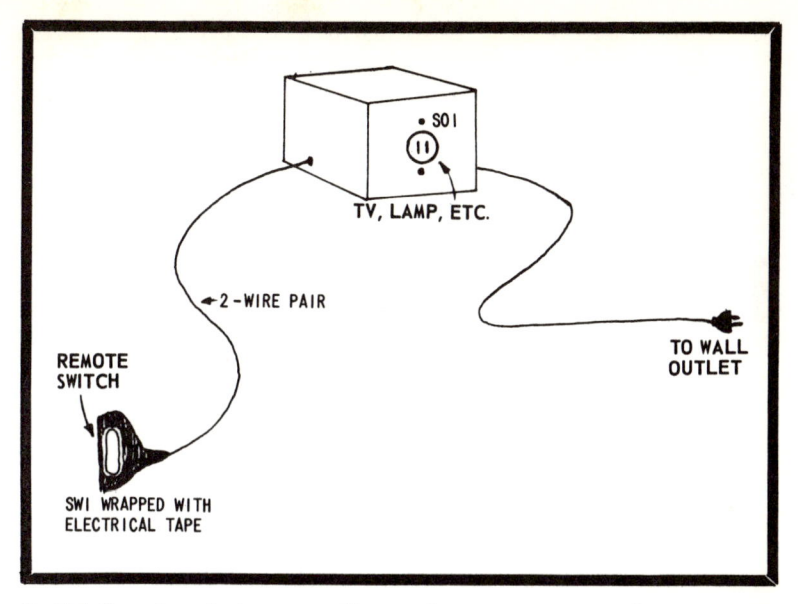

```
                    • SO1
                     (11)
              TV, LAMP, ETC.


         ←2-WIRE PAIR

REMOTE
SWITCH                              TO WALL
                                    OUTLET

SW1 WRAPPED WITH
ELECTRICAL TAPE
```

Build the circuit in a small metal container, such as a Mini-box. The "hot" lines—those in the primary—should all be insulated hookup wire, even if they are inside the box. It is a good idea to cover solder connections with electrical tape just to make sure that you don't fall prey to any "accidental" grounds. AC line cord is a must for the connection to the wall outlet. Place the socket to plug in the controlled appliance in either the side or the top of the box.

The low-voltage transformer secondary leads can be any length you want without worrying about extra insulation or any of the other problems encountered with exposed high-voltage wiring. The connection can be made with ordinary bell wire. Your remote control slide switch is connected in this low-voltage secondary and is insulated as shown in the pictorial with electrical tape. When closed, the switch triggers the relay, which in turn hooks up the appliance. The switch can be either left "free," so that you can hold it in your hand, or it can easily be taped to the underside of a table, chair, or even to the wall, depending, naturally, on what you're using it for.

We have only touched the surface of the sea of possibilities and applications available to you. If you're interested in making your life a little easier, you will find many places where it will be very useful to you. If <u>you</u> can't think of any, surely your wife can!

8
Low-Voltage Closet Wiring

Merely thinking unkind thoughts about that dim closet or storeroom downstairs with that hidden or hard-to-get-at light switch is not going to solve the problem. Here is a circuit that will turn the ceiling light on whenever the door is opened. If the closet is frequently used, you will find this automatic light circuit as indispensible as the one in your refrigerator.

A normally-closed switch is mounted on the door frame where it will come near a magnet placed on the door. The

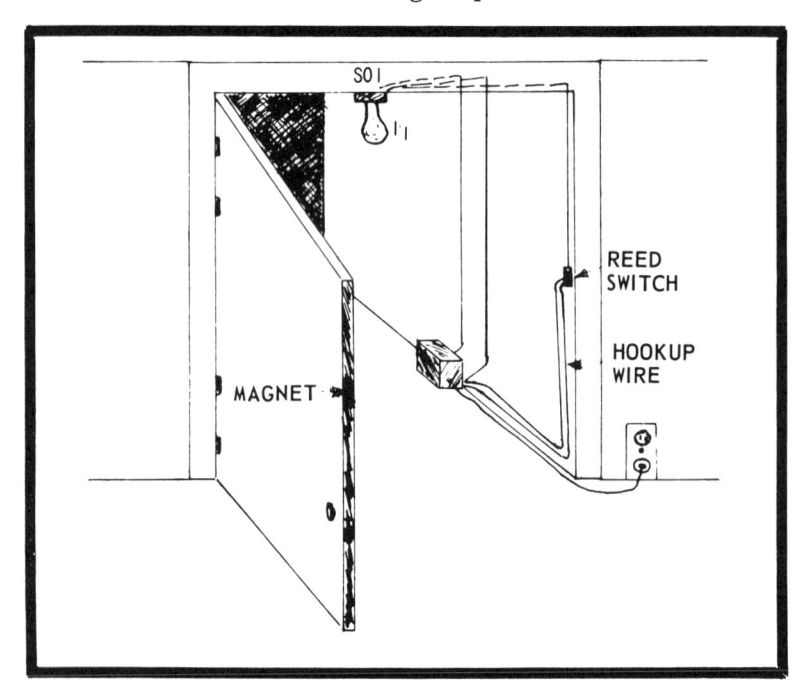

56

PARTS LIST

F1 — 3 amp 117V AC fuse. Littelfuse 3AG
I1 — Household light
J1, 2 — Test jacks. Amphenol 78-1S
PL 1, 2 — Single-prong test plugs. Amphenol 71-1S
PL3 — AC wall plug. Amphenol 61-F11
SO1 — Household light socket
SW1 — Reedswitch, normally-closed
T1 — Alco FR-101
Fuse holder. Buss no. 4405
Length of AC line cord
Length of insulated hook-up wire
Magnet, thin, rectangular, cabinet type
Metal box

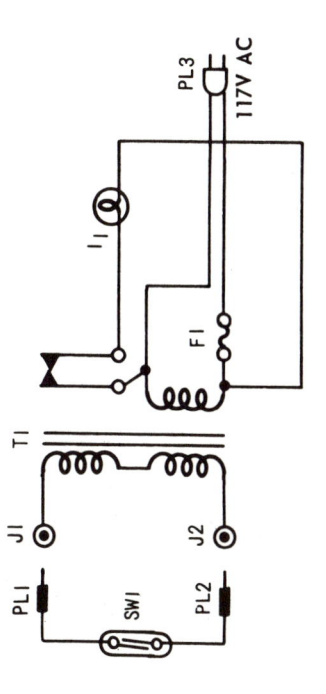

switch is in the secondary circuit of the isolation relay. When near the switch the magnet forces it open, thus insuring that the relay, across which is connected the ceiling light, is open when the door is closed. As the magnet gets farther and farther away from the reed switch, its influence is lost; the relay closes, and the light goes on.

A small box is used to house the relay and fuse. This isolates all the necessary high-voltage connections. The switch leads are low voltage; consequently, they may be left exposed with no hazard. Placed inside the closet, the box must be plugged into the nearest wall outlet with AC line cord. Running the cord under the closet door should create no problem. The holes should be grommetted in the box, however, to protect the insulation on the leads.

If you do not already have a light socket in the closet, follow normal installation procedure with an outlet box, etc. Heavy wire must be used between the box and the socket. Local building codes usually specify the use of conduit for permanent connections. Use of heavy wire, however, is quite safe. Break into the two leads in the outlet box for the switch connections if a fixture already exists. If you take the time to construct this device, you will find yourself saving time and frustration spent groping for the switch in that dark closet. It is simple and it is safe. It speaks for itself.

9 High/Low Soldering Iron Control

During those long evenings spent assembling electrical kits and **projects,** the tip of a soldering iron can really deteriorate. The metals in these tips are subject to corrosive forces all the time, but the condition is noticeable only at higher tem-

PARTS LIST

D1 — 200 PIV, 1½ amp per 100-watt, silicon rectifier (see text)
F1 — 3 amp 117V AC fuse. Littelfuse 3AG
J1, 2 — Test jacks. Amphenol 78-1S
PL1, 2 — Single-prong test plugs. Amphenol 71-1S
PL3 — AC wall plug. Amphenol 61-F11
SO1 — AC female socket. Amphenol 61-MIP-61F
SW1 — Normally-closed leaf-type microswitch
T1 — Alco FR-101
Block of wood for mounting SW1
Fuse holder. Buss no. 4405
Length of AC line cord
Length of insulated hook-up wire

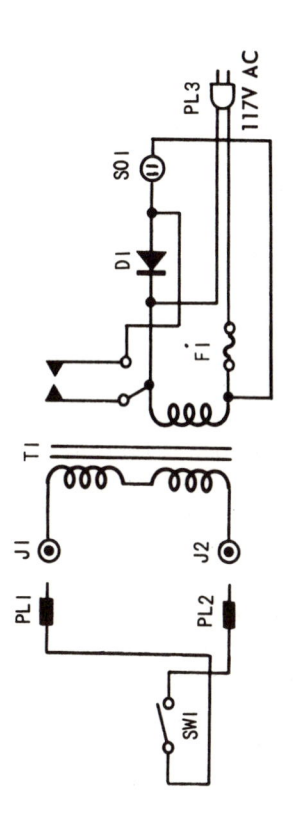

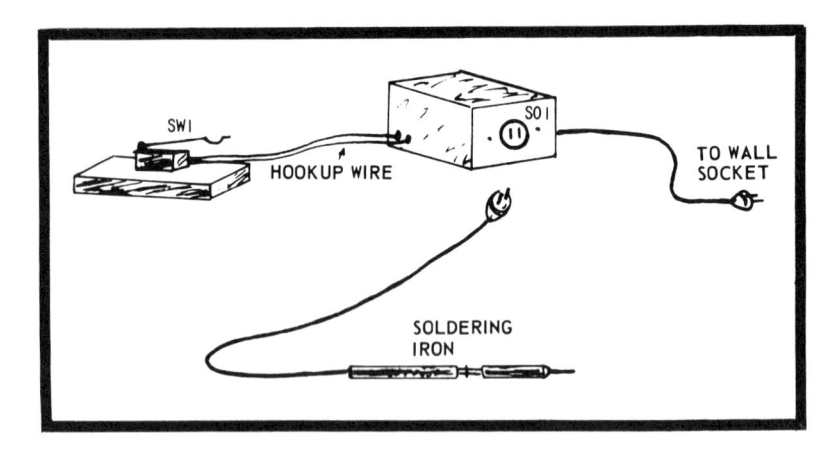

peratures where the reaction rate is greatly accelerated. The hot metal surface, oxidized by the oxygen in the air, takes on a crusty appearance, noticeable especially when the tip becomes loose from time to time. The whole dilemma arises because the iron must be kept at full heat all of the time. If you unplug it when it is not in use, it takes far too long to get it hot again. This project gives you a convenient way to cut the neat in half when the iron is not in use, keeping it at a good temperature for rapid reheating. Probably doubling or tripling the life of your iron tips, this hi-lo switch will make the small investment involved worthwhile in the long run.

To cut the current through the iron in half, a silicon rectifier is used. When the iron is plugged into the socket in the primary of the isolation relay, a closed circuit exists from the 117v AC through the rectifier to the iron. The rectifier passes current in only one direction; consequently, the heat level when the relay is open is one half what it normally is. The rectifier rating, of course, is determined by the wattage of your iron: For every 100 watts the rectifier should be good for 1 1/2 amps.

A normally-closed leaf-type microswitch in the secondary circuit opens and closes the relay. When the switch is closed, so is the relay, and the iron is in a circuit with no current barriers, allowing it to reach full heat. The leaf switch is mounted on a small piece of wood for stability. These switches usually come with an arm that can serve as a holder as shown in the pictorial; if not, you can easily attach one of your own.

The isolation relay, fuse, and socket are mounted in a small metal box. Insulated hook-up wire is used for all connections, including those made inside the box. The wall connection, of course, is AC line cord. Since the secondary voltage is extremely low, these connections are safely left exposed.

Whenever the iron isn't being used temporarily, set it on the arm of the switch, which opens the relay, cutting in the rectifier for "low" operation. Removing the iron from the arm will bring you back to full heat. You needn't think twice about it.

The set-up is easily stored and kept out of the way on your bench. The leads to the switch can be as long as necessary, making it possible to stick the main box underneath the bench if your working space is at a premium.

10 Low-Voltage Lap Counter For Slot Cars

If you have a slot car drag strip in your basement and have enjoyed the excitement of a race with your son once in a while, you probably have gotten to the point where you would like to add a little something extra to the contest. Instead of a race against each other, this can be a race against time. Simulating actual raceway practice, the shortest time for a given number of laps wins. The tedious job of counting the laps, however, can take the fun out of it. For this reason, you will probably be interested in this easy-to-build lap counter.

A normally-open reed switch is connected in the secondary of the isolation relay. These switches are very inexpensive, and can be picked up at any electrical supply store. Two flexible, magnetized metal strips in a glass envelope are brought together when a small magnet is brought into the near vicinity;

PARTS LIST

EC1 — 117V AC electric counter with reset lever
F1 — 3 amp 117V AC fuse. Littelfuse 3AG
J1, 2 — Test jacks. Amphenol 78-1S
PL1, 2 — Single-prong test plugs. Amphenol 71-1S
PL3 — AC wall plug. Amphenol 61-F11
SO1 — AC socket. Amphenol 61-MIP-61F
SW1 — Normally-open reed switch. Gordos MR-107
T1 — Alco FR-101
Cabinet-type rectangular magnet
Fuse holder. Buss no. 4405
Length of AC line cord
Length of insulated hook-up wire
Metal box

in this case, when the slot car with the magnet attached passes by (see pictorial). Because the magnet is near the switch only for a short time, the secondary of the isolation relay is closed only momentarily. This also closes the relay in the primary and the electric counter receives a pulse of current.

The counter is operated on 117v AC and has a reset lever to faciliate re-use. Every time it is connected, the counter advances one number, going no further until it is disconnected and reconnected again. Consequently, everytime the reed switch is forced to close for an instant by the passing magnet, the counter is cut into the circuit just long enough to register one more number.

Enclose the circuit in a small metal box and mount a socket in one side to plug in the counter. All high-voltage connections are made inside this box. For extra safety, a fuse is included in the circuit. The secondary is low-voltage, so exposed leads to the reed switch are perfectly safe.

A small rectangular magnet (the cabinet-type) is attached to the under side of your slot car. The magnet is small enough so as not to hinder the operation of the car. Attach the reed switch to the table near enough to the track to be sensitive to the passing car magnet. The main box may be placed wherever it is convenient, perhaps under the table. It has to be plugged in, so a wall outlet will need to be within cord length.

To put the project into operation, make the necessary connections and plug everything in. With the counter reset to zero, and a stop watch for timing the runs, you'll be all set.

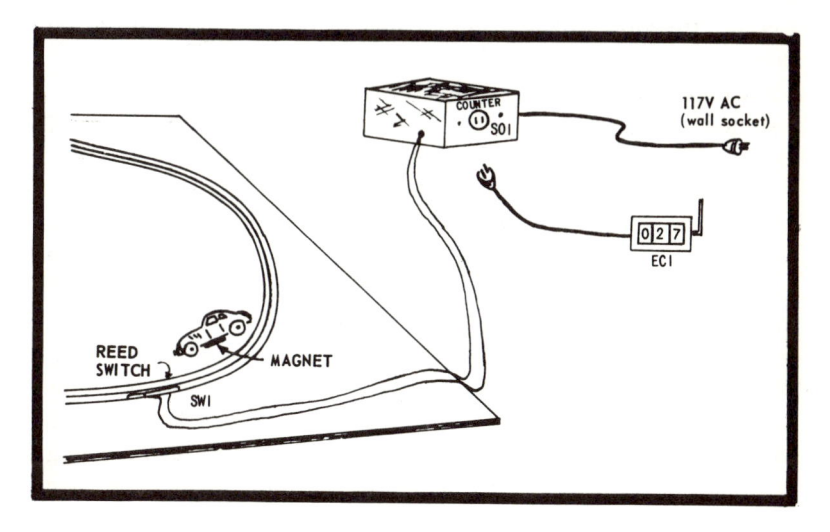

63

11

Effective Home Burglar Alarm

Burglary is all too common in today's society to be ignored. Unfortunately, the normal precautions of keeping the door carefully locked, keeping little money, jewelry and few valuable articles around the house are often not enough. If you are concerned about the possibility of someone forcing his way into your home, endangering not only your possessions but your family as well, then you may be interested in this simple burglar alarm.

In the control leads of the isolation relay are connected a pushbutton or microswitch (used to close the relay), a capacitor, and a key switch (see schematic). The key switch is included so that with a special key you can reset or turn off the alarm, providing the intruder cannot discover the main alarm box.

The capacitor and the secondary coil of the transformer form a series resonant circuit. Unless the reactances of the capacitor and the coil are approximately equal, there will be a sizable voltage drop in the secondary. The circuit is not designed for resonance at 60 hertz (cycles), and besides that, the capacitive reactance of C1 is extremely large at this low frequency. This voltage drop is too large to allow the relay in the primary circuit to close. What is needed is a closed circuit with no current barriers, such as is provided when switch SW2 is closed. It needs to be closed only momentarily—just long enough to snap the relay shut. Now, here is where the catch comes in. The circuit through C1 does carry enough current to keep the relay shut; so as long as SW1 remains

PARTS LIST

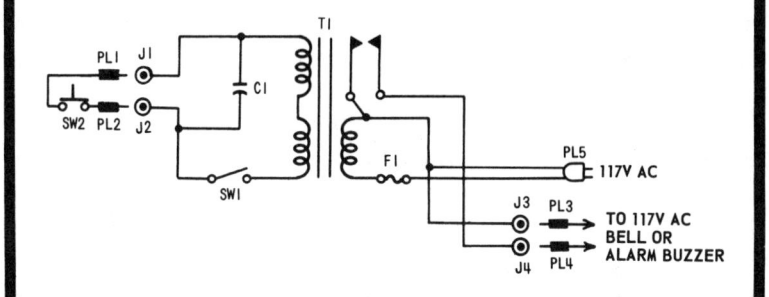

BZ1 — 117V AC bell or buzzer
C1 — 2 mfd
F1 — 3 amp 117V AC fuse. Littelfuse 3AG
J1, 2, 3, 4 — Test jacks. Amphenol 78-1S
PL1, 2, 3, 4 — Single-prong test plugs. Amphenol 71-1S

PL5 — AC wall plug. Amphenol 61-F11
SW1 — Key switch, SPST type
SW2 — Normally-open SPST pushbutton or microswitch (see text)
T1 — Alco FR-101
Fuse holder. Buss no. 4405
Length of AC line cord
Length of insulated hook-up wire

closed, so does the relay—across which is connected the alarm device.

Use a small metal box for construction of the circuit. High-voltage primary connections are all made inside the box. This includes the wall plug connection and the hook-up for the alarm bell or horn, which operates on 117v AC. In the side of the box is mounted the key switch. The capacitor also goes inside. Exposed are only the low-voltage leads to the microswitch, which are safely made with bell wire or zip-cord. They need not be heavy wire, because once the alarm is set off, it does not matter if they are broken or severed.

Microswitches come in a great variety of sizes and forms. Merely brushing against a normally-open leaf switch is enough to set it off. The type of switch that you select will depend entirely upon your use for it. A large, valuable object can be placed on a normally-closed switch. If the object is ever moved, the switch will close and the alarm will go off. A "light touch" leaf switch might prove useful on a door frame. The number of ways in which you can apply these many kinds of switches is unlimited.

Since the secondary is low voltage, the length of the leads to the switch is not too important, although extremely long leads are not recommended. Once the switch is triggered, the switch and the leads to it no longer matter because the capacitor <u>inside</u> <u>the</u> <u>box</u> keeps the alarm on. It is wise, though, to place the main box in some other part of the house so it cannot be unplugged by the intruder. You may want to put the box and alarm in your bedroom with the switch at the front or back door. An alarm at the bedside need not be very loud. Any number of these switches can be placed in parallel in the secondary to protect more than one entrance.

An alarm such as this is always a good thing to have around, even if it is used as protection only occasionally. It is easy and inexpensive to construct and has many practical applications.

12
Rain Detector & Alarm

For those whose only clothes dryer is the sun, an accurate system for detecting the beginning of a rainstorm before it really starts to pour could save the time and trouble of having to re-dry clothes. Another main value of this project is in detecting sump pump failures before it is too late. In the spring, even homes with normal sewer drain locations have problems.

The core of the project is a moisture sensor which consists of a number of parallel—but open—circuits, any of which can be shorted by a drop of water. Whenever one of these is closed, the alarm goes off.

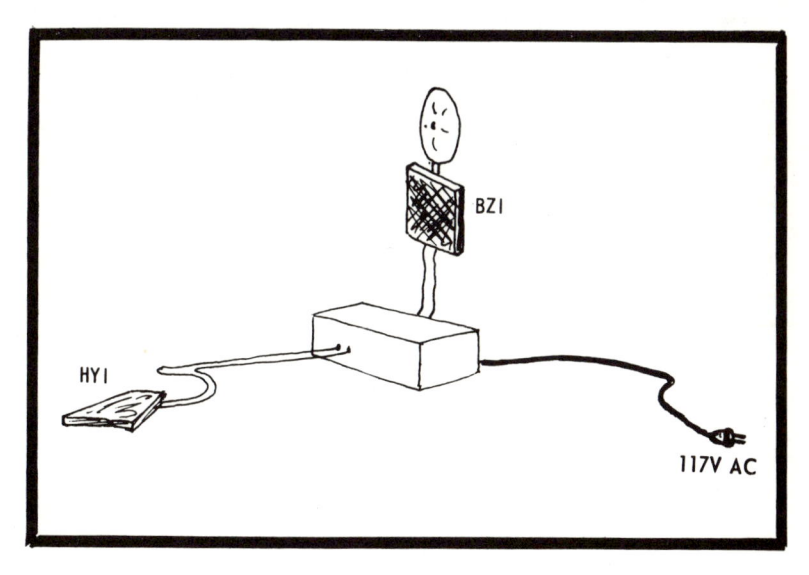

PARTS LIST

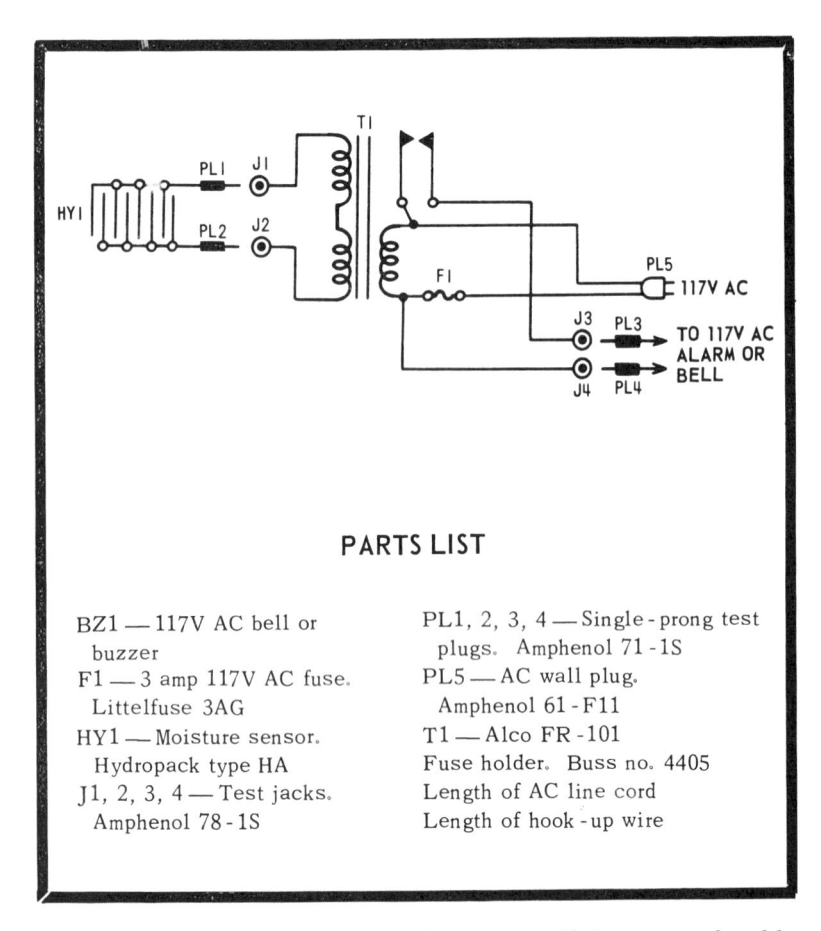

BZ1 — 117V AC bell or buzzer
F1 — 3 amp 117V AC fuse. Littelfuse 3AG
HY1 — Moisture sensor. Hydropack type HA
J1, 2, 3, 4 — Test jacks. Amphenol 78-1S
PL1, 2, 3, 4 — Single-prong test plugs. Amphenol 71-1S
PL5 — AC wall plug. Amphenol 61-F11
T1 — Alco FR-101
Fuse holder. Buss no. 4405
Length of AC line cord
Length of hook-up wire

The components are mounted in a small box, preferably metal. The high-voltage connections to the wall and to the 117v AC alarm bell or buzzer must be made inside this box. The control leads of the isolation relay carry low voltages and are safely left exposed.

The box and alarm are left inside the house and the sensor placed outside for detecting rain; lead length is of little consequence. For detecting water in the basement due to an overflowing sump, place the sensor on the floor near the sump.

13

Shooting Gallery Game

Kid's parties are always dominated by potentially dangerous contests of skill; darts for instance. Here is another fiercely competitive game that will keep the children occupied for hours, but which is completely safe. A pulse of light from a modified toy gun strikes a photocell placed in a homemade target to trigger an electric counter, thus recording the number of hits.

The gun can be fashioned out of a piece of wood, or if you have an old broken toy gun lying around the house, that would be suitable, too. Placed at the end of the barrel is a parabolic reflector from a flashlight, along with a light bulb and contacts. The battery can be put in a small enclosure attached to the butt of the gun. Mounted in the traditional place, the trigger is a normally-open pushbutton switch—leaf type or slide. Press the trigger to shoot a directional pulse of light; aim to get a bullseye.

The target consists of two pieces of wood, fastened with hinges to facilitate simple, flat storage (see pictorial). Draw a target on the vertical piece, drilling or cutting a hole in the bullseye position for the photocell.

The photocell is the resistive type. When hit by a beam of light from the gun, it passes enough current to close the isolation relay, which in turn connects the electric counter. Every pulse of current advances the counter one number.

A small metal box with a socket mounted in the side is perfect for housing the relay. While AC line cord is mandatory for the line running to the wall outlet, the leads to the photocell need be only bell wire. The transformer steps down the

PARTS LIST

B1 — 1½V DC
EC1 — 117V AC electric counter with reset lever
F1 — 3 amp 117V AC fuse. Littelfuse 3AG
I1 — 1½V DC bulb
J1, 2 — Test jacks. Amphenol 78-1S
PC1 — Resistive-type photocell. Lafayette 19 H2101
PL1, 2 — Single-prong test plugs. Amphenol 71-1S
PL3 — AC wall plug. Amphenol 61-F11
SO1 — 117V AC socket. Amphenol 61-MIP-61F
SW1 — SPST pushbutton switch

T1 — Alco FR-101
Fuse holder. Buss no. 4405
Hinges for construction of bulls-eye target
Homemade bulls-eye target

Length of AC line cord
Length of insulated hook-up wire
Reflector for flashlite
Toy gun

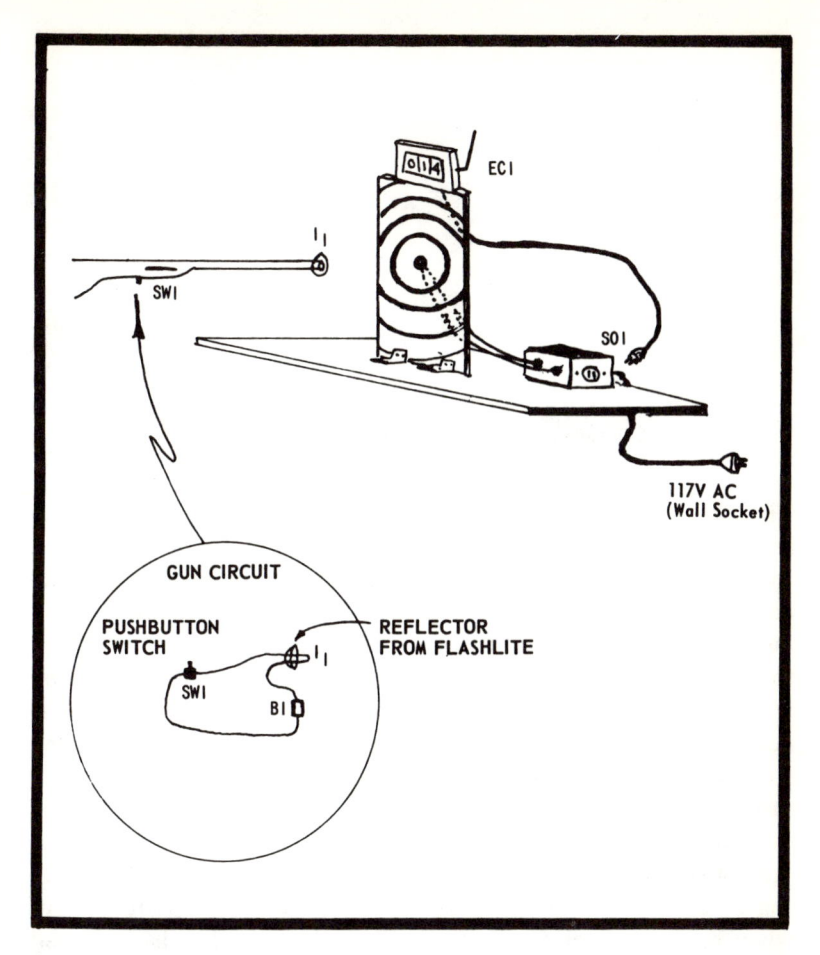

voltage. The 117v AC counter is plugged into the socket in the box and the box is plugged into the wall.

Worry no longer about party or passtime games with sharp projectiles, pins, or other possibly dangerous objects. A beam of light can't hurt anyone, and there is absolutely no chance of shock from exposed leads because of the low voltage. This is the modern version—the "laser"-type version—of the old fashion pellet or cap gun.

14
Audible Continuity Tester

The first test that you make if an electrical device is not working is often a test for continuity to ground at certain points in the circuit. Usually you do this with an ohmmeter, or some equivalent. If the meter reads infinite resistance, and there's supposed to be a partially open line to ground, you know where to look for the trouble. There is, however, a distinct disadvantage to this type of set up. Inserting the probes into the set and hitting just the right component or location for the continuity reading is often a touchy process which requires your full attention. While holding the probes in the right place you have to look at the meter. How often have you shifted your attention from the circuit to the meter and found that the reading on the meter means that you have found your trouble; but when you turn back to the probes you find that one has slipped from the terminal to a nearby capacitor. You might have tried it again only to find that your contact with the probe wasn't any good. You can solve the problem by using this audible continuity tester. Keeping your eyes on the probes to insure their correct positions, you will hear in the earphone a tone if there is continuity, silence if there is not.

The circuit is simplicity itself. The components consist of only a capacitor, the isolation relay, a magnetic earphone, two probes and the other electrical paraphenalia such as insulated hookup wire, etc.

Build the circuit in a Minibox or any small metal container big enough for the relay. Since the probes are connected in the secondary circuit of the transformer where the voltage is quite low, you need not take any special precautions to guard

PARTS LIST

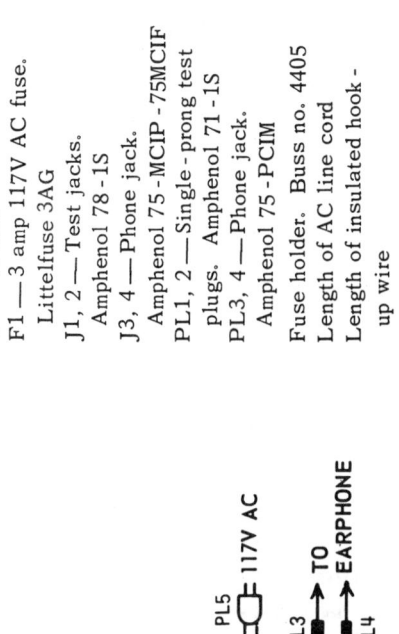

C1 — .3 mfd
F1 — 3 amp 117V AC fuse. Littelfuse 3AG
J1, 2 — Test jacks. Amphenol 78-1S
J3, 4 — Phone jack. Amphenol 75-MCIP-75MCIF
PL1, 2 — Single-prong test plugs. Amphenol 71-1S
PL3, 4 — Phone jack. Amphenol 75-PCIM
Fuse holder. Buss no. 4405
Length of AC line cord
Length of insulated hook-up wire
Magnetic-type earphone
PL5 — AC wall plug. Amphenol 61-F11
T1 — Alco FR-101
2 Probes

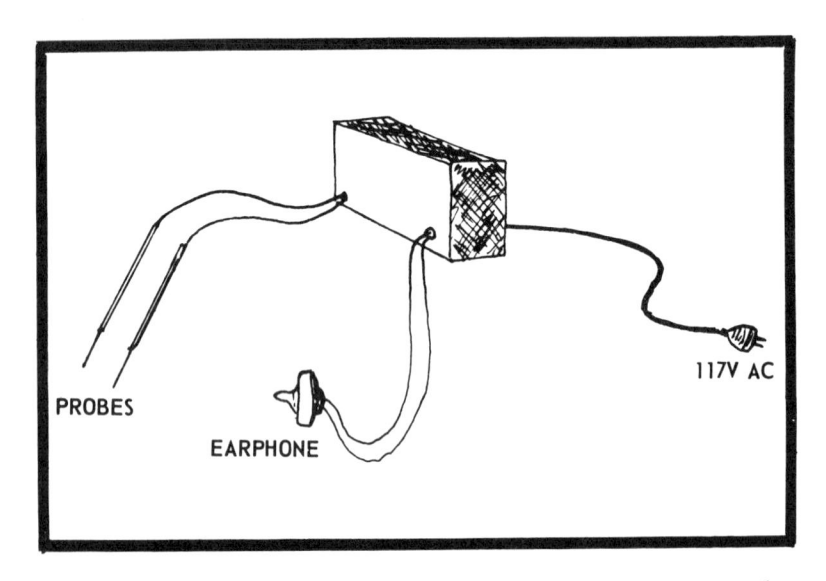

[Diagram: metal box with PROBES on left, EARPHONE below, and 117V AC plug on right]

against shock. It is important to keep in mind, however, that 117v AC is present in the earphone circuit. This means careful insulation of the leads and earphone connections. Be sure to make all connections inside the metal box. If you follow these tips, you will have a circuit that is both safe to operate and to handle. The line to the wall socket is of course AC line cord. A fuse is included here to give you added safety.

The junk that you have lying around the house is all you need to fashion yourself a set of professional looking probes. Use the wire that is already connected in the secondary for the probes. Or, if you happen to have some old narrow glass tubing and an alcohol lamp, heat the glass tubes around the wires at the end to make two excellent insulators. Actually, any dielectric will do the trick. Even wrapping tape around the wires will work. Now you're set to enjoy your new-found convenience. Build this project and you need not worry about misplaced probes while you're testing.

15

Automatic On/Off Precision Drill Switch

When the light in your basement is burning brightly at two o'clock in the morning and you're trying for the 29th time to get a perfect start into a small piece of metal with your drill press, it is time to go to the precise switching method suggested in this project. Seriously, it is very difficult to get an accurate start with an ordinary drill press. Most of the time, when turning at high speed, there is a slight wobble at the point of the drill. As a result, it is impossible to see exactly where you're putting the point down. Very often, a slow start helps a great deal. This means turning the drill on for short intervals of time, which forces you to divert part of your attention from your work to the switch. This can be a problem, too. Using this automatic electrical switching method, your chances of being disappointed by an off-center or ill-fitting hole are greatly reduced. The speed with which you start, the exact location, etc., are completely under your control. All you have to do is touch the drill point to whatever your drilling and the drill press will go into motion.

Construction of this particular project is no different than most of the other projects in this book. Following the usual pattern, a small metal box is used to house the circuit, an especially good idea here because the conditions in a basement or workshop where the drill press is usually kept are very often damp and humid, necessitating extra protection against accidental grounding, etc. The connection to the wall is of course AC line cord. Inside the box, all connections should also be made with insulated hookup wire for added safety. To plug the drill in, a socket is installed in the box.

INSULATE FROM GROUND

SOLDER TO METALLIC BASE

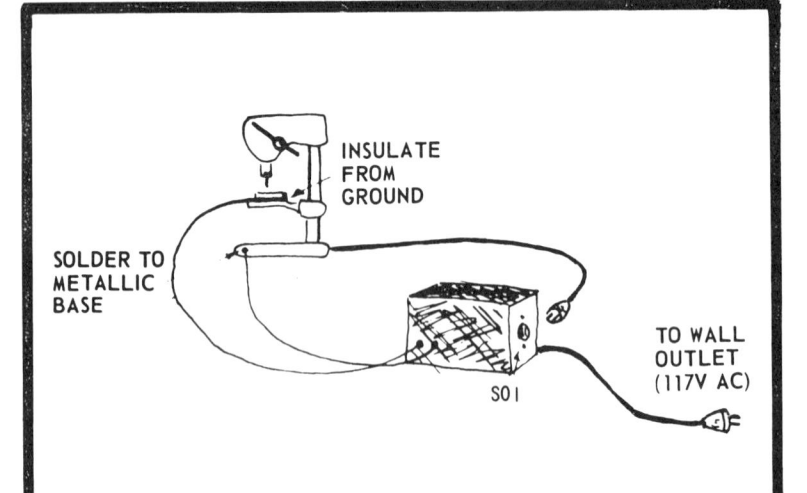

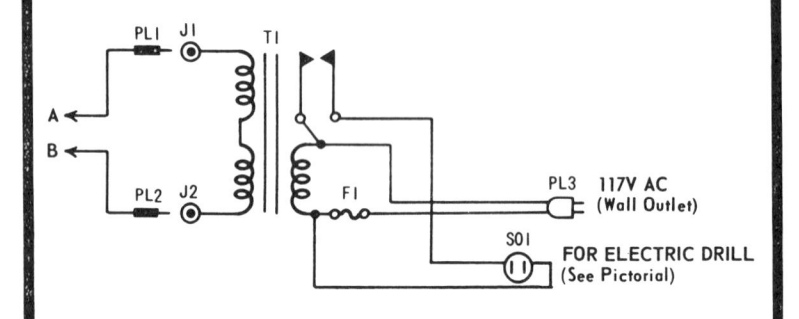

TO WALL OUTLET (117V AC)

SO1

PL1 J1 T1
A ←
B ←
PL2 J2   F1   PL3 117V AC (Wall Outlet)
                       SO1 FOR ELECTRIC DRILL (See Pictorial)

PARTS LIST

F1 — 3 amp 117V AC fuse.
 Littelfuse 3AG
J1, 2 — Test jacks.
 Amphenol 78-1S
PL1, 2 — Single-prong test plugs. Amphenol 71-1S
PL3 — AC wall plug.
 Amphenol 61-F11

SO1 — 117V AC socket.
 Amphenol 61-MIP-61F
T1 — Alco FR-101
Fuse holder. Buss no. 4405
Length of AC line cord
Length of insulated hook-up wire

Connect the drill as shown in the pictorial. The insulated wires that compose the secondary of the isolation relay are run from the box to the drill and the metal being drilled. One wire, and it doesn't matter which one of the two, A or B, is soldered to the metallic base of the drill press. Since the voltage in the secondary is very low, you needn't worry about leaving this connection exposed. Covering it with tape just to keep it out of the way is not a bad idea, though. Assuming that the press is on a wooden bench, the other wire goes to the metal being drilled. This piece of metal must be insulated from both the drill and from ground. Use a thin piece of wood for this purpose. The bare end of the wire can be placed between the wood and the piece of metal. Although soldering is not necessary here, you should make sure you have a good connection.

The circuit operates very simply. Connected across the relay in the circuit, as you can see from the schematic, is the drill (or whatever is plugged into the socket). This circuit will remain open as long as the secondary is open. When any part of the drill press touches the piece being drilled, the secondary is completed, which in turn closes the relay and turns on the drill. Break the connection and the drill will go off. Very simple.

16
Remote On/Off Motor Controller

As a rule, wall outlets are not put out where they can be seen. If they were, they would detract from the appearance of the room. When you hide an outlet where it is not too conspicuous, you also make it a lot harder to get at in order to plug things in. How many times have you come away from a scuffle with one of these hard-to-get-at sockets with a bumped head,

PARTS LIST

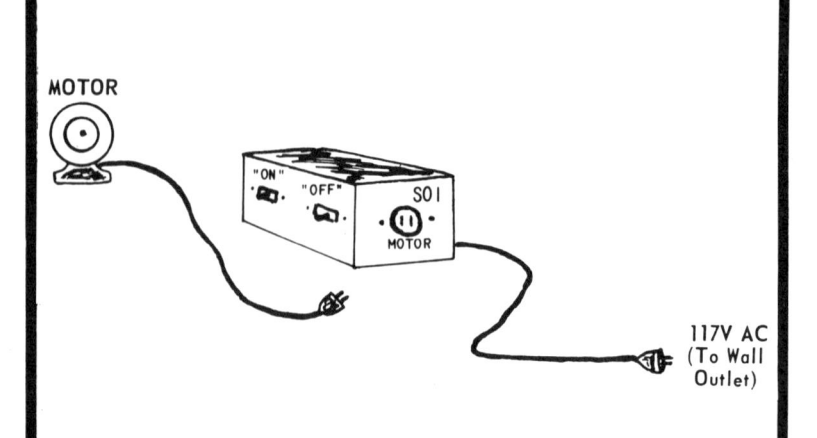

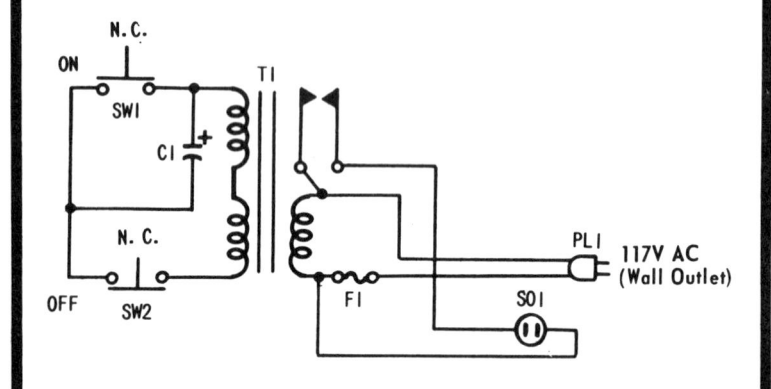

C1 — 2 mfd, 150 WVDC electrolytic
F1 — 3 amp 117V AC fuse. Littelfuse 3AG
PL1 — AC wall plug. Amphenol 61-F11
SO1 — 117V AC socket. Amphenol 61-MIP-61F

SW1 — Normally-open SPST pushbutton
SW2 — Normally-closed SPST pushbutton
T1 — Alco FR-101
Fuse holder. Buss no. 4405
Length of AC line cord

a sore back, and dusty clothes, saying to yourself over and over again that you were going to find a better and more convenient way to do it next time? If you happen to be in the market for the solution to such a problem, you're in luck, because we just happen to have one here. Using it, you can have the controls for the motor, or whatever, right at your fingertips. You won't have to get down on hands and knees and just about kill yourself trying to plug the device in. This is especially useful for those electrical devices that don't have an on-off switch.

The project consists mainly of two SPST pushbutton switches, a capacitor, and the isolation relay. Pushbutton switch SW1 is normally open. The other switch, SW2, is normally closed. The secondary circuit for the isolation relay is never closed because of the capacitor, which can pass AC current. The capacitor does inhibit current, weakening it enough so that it is not capable of closing the relay by itself. When you close SW1, the circuit is clear, and the full amount of current is allowed to flow, causing the relay to snap shut. When SW1 opens, which is immediately after you release it, the relay is kept shut by the current that was present initially in the secondary. By pressing SW2 you cut off all current in the secondary and the relay opens.

One especially attractive feature of the circuit is that you don't have to come in contact with high voltage at all, except when you initially plug it in. This is a lot safer than having to handle and insert a plug every time you want to turn something on.

You can mount the entire circuit (except for the plug) inside a small metal box. The two switches can be mounted side by side, each one being labeled as to its function. Make all connections inside the box with insulated hookup wire. The line to the wall outlet, of course, is AC line cord. The box's appearance can be improved by covering it with attractive paper, formica, or even finished wood. Another idea is to sink the box into the desk, making it flush with the surface.

Since the circuit has a socket in it, any appliance or motor that does not have its own switching mechanism can be plugged in and operated with convenience and safety. So it's versatile, too! Perhaps you have a soldering iron that has to be plugged in to a hard-to-reach socket every time you want to use it. Turn it on and off with the push of a button. Another use that comes to mind is aquarium air pumps. Frequently these pumps must be turned on or off at least twice a day. This means using the plug every time. It is not only irksome, but dangerous, because of dampness and wet hands.

17
Electric Candle That Lights With A Match

In the beginning the Master Workman created the heavens and the earth. The earth was without form and darkness was upon the face of the deep. And it was said, "Let there be light": and there was light. And the Creator saw that the light was good: and He separated the light from the darkness.

Man, blessed by the other labors of the first six days, has accumulated a vast knowledge of electronics and practical electrical projects. One such project is to turn on an electric candle with a match as though the candle were made of wax and string. You can always depend on this to be an attention getter at a party. If it is done in a dimly lit room, the lighting of the candle can be not only mystifying, but eery.

The main parts for the project are a resistive photocell, an electric candle, and the isolation relay. Since the photocell is resistive, the secondary is not open and current does flow. This is necessary because the current generated by the photocell when it is illuminated is not enough to close the relay by itself. In other words, the sum of the two currents is needed to close the relay. When the relay is closed, of course, t the candle goes on.

If you don't like the idea of the candle, you can choose any 117v AC light source that can be plugged in, since the circuit

PARTS LIST

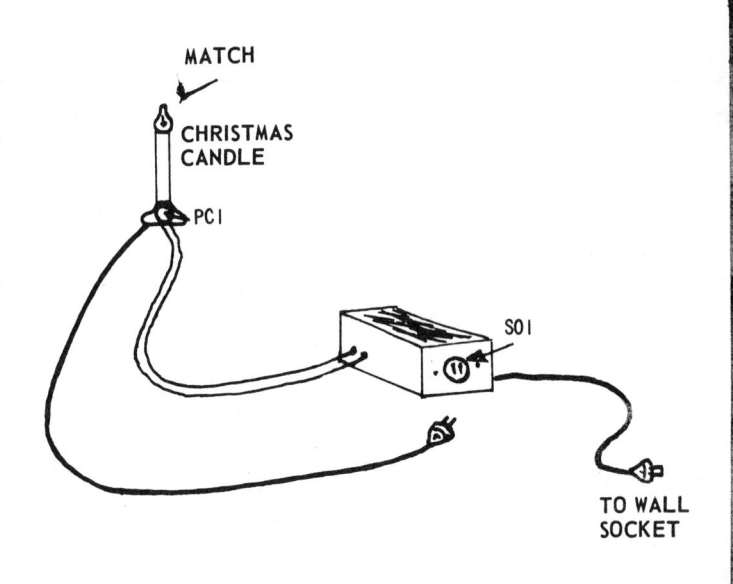

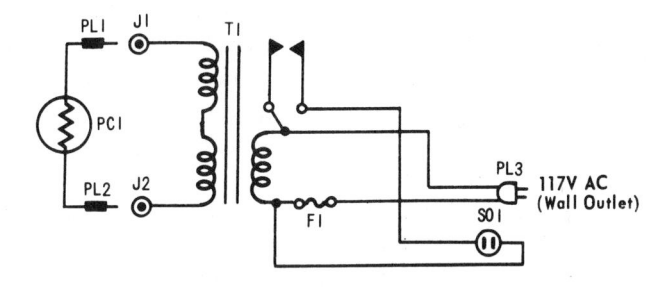

F1 — 3 amp 117V AC fuse.
  Littelfuse 3AG
J1, 2 — Test jacks.
  Amphenol 78-1S
PC1 — Resistive photocell.
  Lafayette 19H2101
PL1, 2 — Single-prong test
  plugs. Amphenol 71-1S
PL3 — AC wall plug.
  Amphenol 61-F11

SO1 — AC socket.
  Amphenol 61-MIP-61F
T1 — Alco FR-101
Electric candle
Fuse holder. Buss no. 4405
Length of AC line cord
Length of insulated hook-
  up wire

contains a socket instead of a permanent connection. An electric lamp will work just as well as an electric candle. The relay is closed only when light is striking the photocell, and it will open unless there is another source of light present after the match is blown out. Once the candle goes on, it keeps the photocell going by its own light. The same would be true of the lamp. Even a TV might work.

Mount the isolation relay, socket, and fuse in a small metal box. The photocell, because of the secondary coil's low voltage, can be connected with ordinary insulated hookup wire or bell wire. All high-voltage connections must be made in the box, and the wire to the wall outlet must be AC line cord.

You can put the photocell just about anywhere. If you're using an electric candle, the base of the candle might be a good place. Remember that it doesn't take much light to set it off. In a room that is normally lit, it would light immediately. As a rule, a darkened room is a necessity. The distance between the light and the photocell makes a difference, too. For a candle it might have to be relatively close, but with a lamp you may be able to have it all the way across the room. This is the only way it will work if you want the light to stay on in a dark room after the match is blown out. The best arrangement for your own situation is up to you. You can hide the metal box anywhere within the limits of the length of the wire to the photocell and the wall outlet.

You are ready now to mystify your friends and family by lighting an electric candle with a match. The real trick is that it will stay on after the match is blown out. You'll have them confused until some wise guy finds one of the leads or the box, and even then they still might not know how it works. This thing may be good for a seance also. Maybe that's how they do it!!

18
Add-A-Switch Using Exposed Wiring

When you get home from work and pull into the garage, it is convenient to be able to see where you're going. Frequently the switch for the overhead garage light is nowhere near your car door and you're lucky if you get out with only bumped shins and dusty clothes. You can solve your problem by installing a low-voltage switch for the light where you need it. In fact, using this remote control method you can turn any appliance on or off from any location in the house. Another idea is having a switch near your bed for turning the ceiling light on and off.

The project consists of only an SPST slide switch and the isolation relay. The slide switch, when closed, causes the relay to close and the light or 117-volt appliance to go on.

The isolation relay is small enough to fit inside most light fixtures or the fixture and outlet box. If necessary, another outlet box can be added to accommodate the transformer and fuse, so it will remain out of sight above the ceiling. If appearance does not matter a great deal, use a small metal box and mount it next to the fixture on the ceiling.

Because of the low voltage in the secondary, the remote control switch can be wired with ordinary bell wire or zip-cord. If you want to conceal the wires without having to put them behind the wall, you can buy tape with wire contained right in it, enabling you to fasten the wire down and out of your way. If you decide to use the switch in the basement or garage, concealing the box and wiring isn't too important.

Basement stairs—if there is no way of turning the light on at the top—can be extremely dangerous because of poor lighting; therefore, an extra switch for the basement fixture adds

PARTS LIST

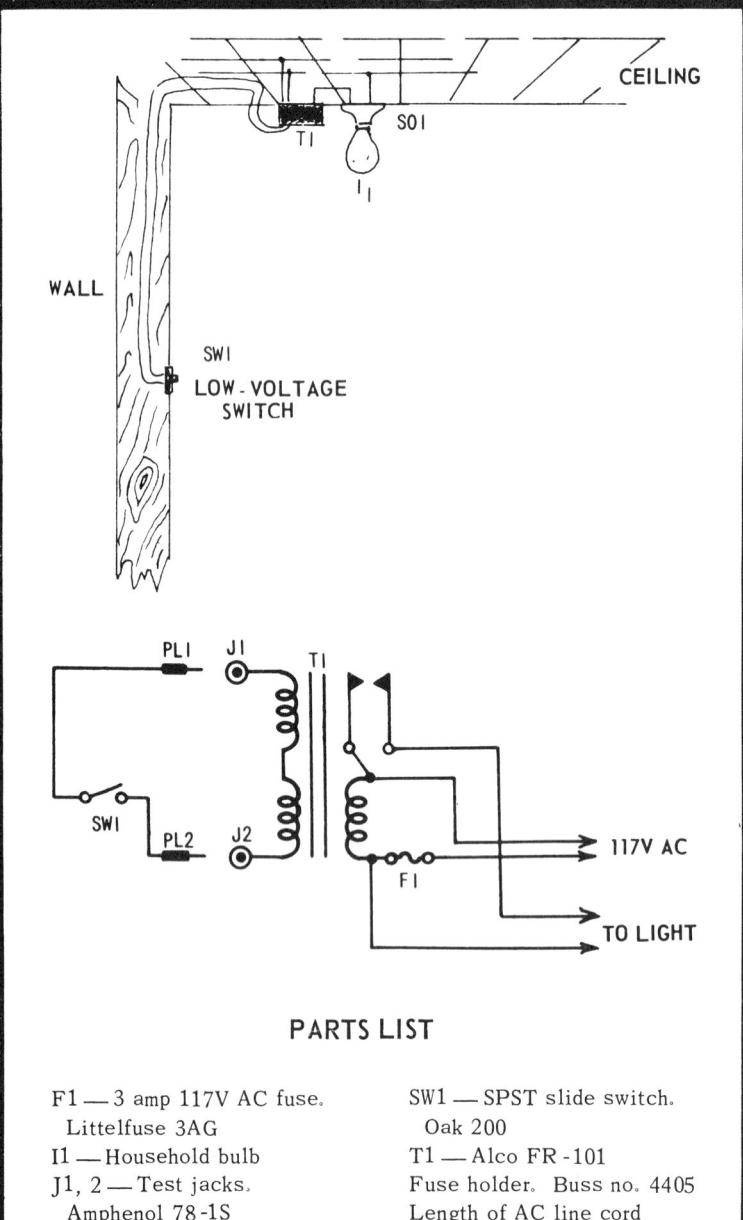

F1 — 3 amp 117V AC fuse.
Littelfuse 3AG
I1 — Household bulb
J1, 2 — Test jacks.
Amphenol 78-1S
PL1, 2 — Single-prong test
plugs. Amphenol 71-1S
SO1 — Porcelain household
light socket

SW1 — SPST slide switch.
Oak 200
T1 — Alco FR-101
Fuse holder. Buss no. 4405
Length of AC line cord
Length of insulated hook-
up wire

a measure of safety for you and your family. Also, basement workshops are often lighted by a large fluorescent fixture which has to be plugged in as a means of turning it on or off. A low-voltage switch would be a welcome change from fumbling around in the dark with 117 volts. Plugging in a soldering iron every time you want to use it can be irksome, too. Use this switch instead. In a place like a utility room where hands are often wet, an increased margin of safety is needed. A low-voltage switch is ideal. And—as an ultimate convenience —you could even turn the morning coffee on while you're still in bed by mounting a switch that controls a wall outlet in the kitchen!

19
No-Load Safety Protector for DC Supplies

In most high-voltage power supplies there is a voltage-dropping resistor to prevent drastic voltage drops from no-load to loaded conditions. Also, supplies with both positive and negative outputs always need a resistor to separate the two potentials. If a supply such as this is turned on without a load, it will take quite a while to clear the air of the smell and smoke of the burning resistor and secondary of your transformer. Suggested here is a safeguard mechanism to prevent you—no matter how hard you may try—from turning the supply on with no load.

The secondary of the isolation relay and the line from the supply to the load are linked by a single-pole double-throw switch. The switch must be closed to set the relay in motion. When the switch is closed, the load is also connected. Due to a slight mechanical delay in the relay, the supply is turned on a few moments later. No matter what, the load is con-

PARTS LIST

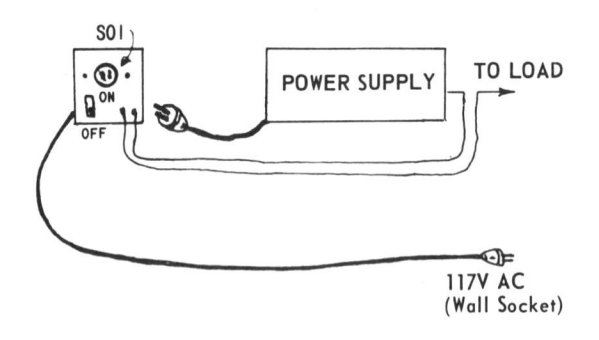

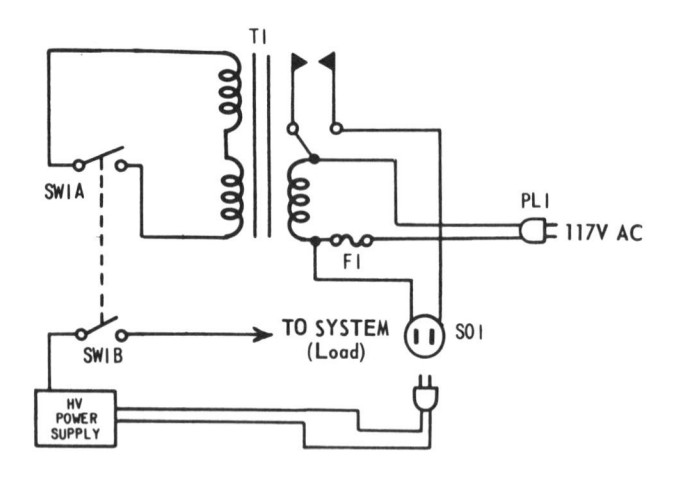

F1 — 3 amp 117V AC fuse.
Littelfuse 3AG
PL1 — AC wall plug.
Amphenol 61-F11
SO1 — AC socket.
Amphenol 61 -MIP -61F
SW1 — DPST slide switch.
Oak 200

T1 — Alco FR -101
Fuse holder. Buss no. 4405
Length of AC line cord
with plug
Length of insulated hook-up wire
Metal box

nected first. The supply is connected across the relay (see schematic).

Switch SW1 and the socket for plugging in the power supply should be mounted in one side of a small metal box. Inside the box go the fuse and isolation relay. The leads running from the supply to the box to the load must be capable of carrying the voltage and current demanded by the load. If you follow these suggestions you will have no trouble. Now all that has to be done is to be sure that there is a load connected to the control box.

20
Magic Wand Lamp Switch

When you are thinking of throwing a party, are you ever plagued by the dilemma of needing something to keep the action going after the card tricks, jokes, and records have just about been exhausted as attention getters? Here is a novelty that both you and your friends can get a lot of fun out of. Although it is a little unusual, it can be put to some practical uses. With a flourish of an easily constructed magic wand (because of modifications in the light circuit) you can magically turn off the light. Actually, anything that is magnetized will do the trick. After the oo's and the aah's have subsided, if you used something as inconspicuous as a magnetized ring, your friends' efforts will be frustrated in trying to duplicate the magical happening.

The project consists mainly of two reed switches, a lamp (which you probably have in copious quantities around your house), and the transformer-relay system. The switching circuit is necessarily different and a little more complex than most of the other projects in the book. Interacting with a capacitor (C1), the two reed switches serve to turn the lamp on

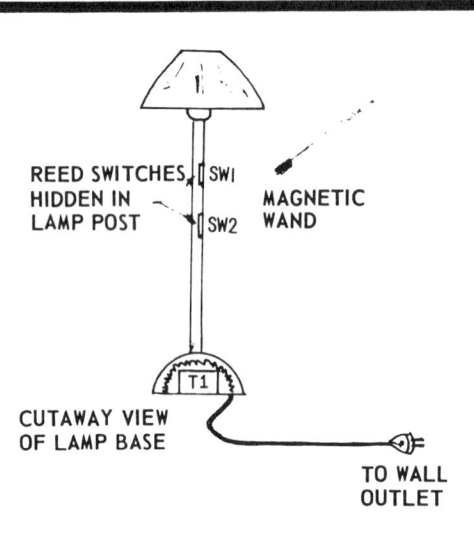

REED SWITCHES [SW1
HIDDEN IN
LAMP POST [SW2

MAGNETIC WAND

T1

CUTAWAY VIEW
OF LAMP BASE

TO WALL
OUTLET

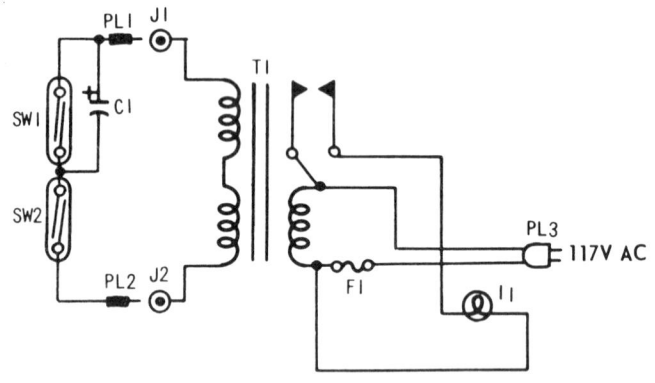

PL3 — 117V AC

PARTS LIST

C1 — .2 mfd, 250 WVDC electrolytic
F1 — 3 amp 117V AC fuse. Littelfuse 3AG
I1 — Household bulb, in lamp
J1, 2 — Test jacks. Amphenol 78-1S
PL1, 2 — Single-prong test plugs. Amphenol 71-1S
PL3 — AC wall plug. Amphenol 61-F11
SW1, 2 — Normally-open reed switch. Gordos MR-107
T1 — Alco FR-101
Fuse holder. Buss no. 4405
Length of AC line cord
Magnetic wand

and off. These switches are nothing more than very thin, flexible pieces of metal that can easily be deflected by a magnet placed anywhere near them. When the switch is affected by a magnet, the circuit will close or open, depending on whether it is normally open or closed. In this circuit, SW1 is normally open and is the "on" switch, and SW2 is normally closed and is the off switch.

After looking at the schematic you may wonder why it works. It is not readily apparent without an explanation. Ordinary reed switches will stay closed or open only momentarily, in other words, when a magnet is in the near vicinity. Ordinarily then, the relay would snap open whenever the magnet is taken away and the circuit opens. The key here is the capacitor that is connected across reed switch SW1. The voltage induced in the secondary circuit is AC. Because of the capacitor's ability to pass AC current, current will flow in the secondary. Not as much current will flow, however, if the capacitor is there than if it is not. In this case, the current is not quite great enough to close the relay, an additional boost being needed. SW1, when it is closed, completes a circuit with no capacitor in it, and the current is great enough to close the relay. When SW1 is opened, the capacitor current, which is still present, keeps the relay closed until the normally-closed SW2 is opened, cutting off all current in the secondary. So, if you touch the first switch, the light goes on; touch the other and the light goes off. Very simple, very convenient, and very safe because of the low voltage in the secondary.

Mount the transformer and fuse in a small metal box. All high-voltage connections must be made inside this box with insulated hookup wire. Ordinary bell wire can be used for connections between the switches, transformer, and capacitor. It also might be wise to insulate all connections with electrician's tape. For convenience, place the capacitor near SW1.

The box and switches can be placed almost anywhere for the desired effect. One idea is to place the box in the base of a lamp and hide the switches in the lamp post. Actually, the unit and switches can be placed anywhere in the vicinity, someplace where they can't be immediately found, of course. In the arm of a chair, in the wall, or right under the floor are other possibilities.

You can make a magnetic magic wand very easily from a

kite stick and a very small magnet, an item which probably can be found around the house. The stick can be painted or covered with paper. Ideally, the magnet should be concealed, also. A magnetized ring, as an alternative, should really confuse your victims.

Now you are ready to dazzle and befuddle your friends and peers by brandishing a magic wand, saying a few abra-kadabra's, and turning off a light, apparently by magic. Almost anything in your house could be controlled this way, if properly wired. Turn on your TV by tapping the arm of a chair. Place a pair of switches on your front porch to enable you to turn on the inside lights before you go in. The possibilities go on and on. Besides its being the highlight of a party, your kids would get a big kick out of playing with it and trying to figure it out. Trade a minimum of trouble and time for a lot of fun!

21
"Ring" That Opens An Electric Door Lock

Seeing small children playing in the back yard or patio has been a universal joy throughout the ages. Whether they be your own or a neighbor's, their joyful cries and playful frolicing can be heard wherever there's mud and lots of room to run around. If you have a swing set in your yard, you probably own the converging point for kids from blocks around. Despite the many pleasures they bring, neighborhood kids can be a nuisance, especially if they bring their dogs with them. Ruined flower beds, unbearable noise, and certain other things that dogs leave around are among the undesirable aftereffects. If you have a fence around your yard and still have a problem, your gate does not keep them out, obviously. A

RING
METAL TRIM

TO WALL SOCKET

ELECTRIC DOORLATCH

T1

TO GATE
(See Pictorial)

F1

PL1

117V AC

TO ELECTRIC DOOR LATCH

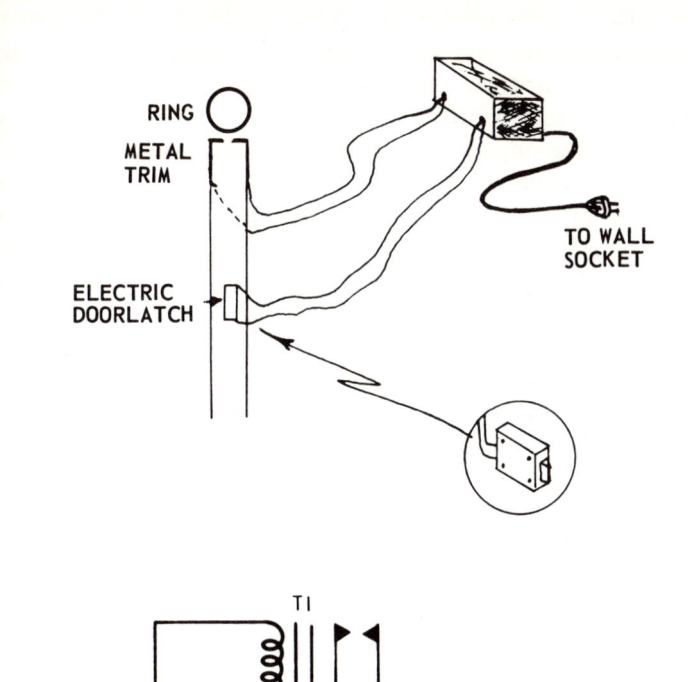

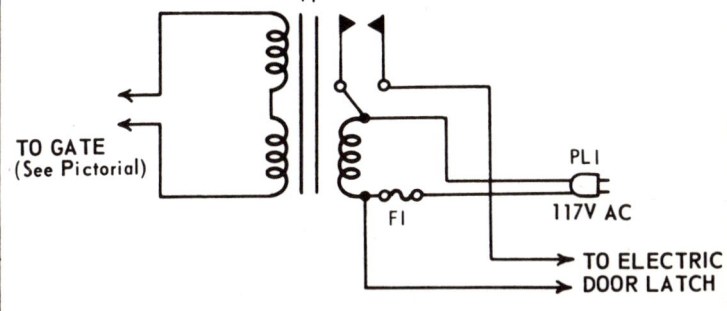

PARTS LIST

F1 — 3 amp 117V AC fuse. Littelfuse 3AG
PL1 — AC wall plug. Amphenol 61-F11
T1 — Alco FR-101
Fuse holder. Buss no. 4405
Length of AC line outdoor-type line cord

Length of insulated hook-up wire, outdoor type
Metal ring
117V AC electric doorlatch
Sections of metal trim
Weatherproofed metal box

padlock is out of the question because then even you would have a little trouble getting in. Your problem may be solved with this simple method of locking and opening your gate. By the touch of the ordinary ring on your finger, you unlock the electric doorlatch, while anyone else, not knowing the trick, will remain outside, or in, unless he wants to hop the fence.

The project consists mainly of an electric doorlatch, the isolation relay, and two pieces of metal trim. The circuit operates very simply. Complete the secondary and the isolation relay closes, connecting the electric doorlatch, so that the gate can be pushed open. As long as the secondary stays open, the gate remains locked. The switch in the secondary is the 2-piece metal trim and the ring. If the ring touches both halves of the rim at the same time, the circuit is closed. Almost any two pieces of metal or foil, as long as they are inconspicuous, can be used.

Don't be concerned about getting a shock when you touch the rim with the ring, because you won't get one. The voltage, which is greatly stepped down from 117 volts AC, is not enough to give you a jolt. Only a metal object will trigger the relay because your fingers do not conduct enough current to activate it.

You can mount the isolation relay and fuse inside a small, weather-proofed box. Be sure to make all high-voltage connections inside this box; that is, the hookup between the transformer and the cord to the wall and the hookup to the cord for the doorlatch. Because this circuit will be out in the weather, it must be durable. Consequently, the AC line cord to the wall socket had better be the outdoor type. For the connections to the trim, although the voltage is low, outdoor-type insulated hookup wire is a wise choice. The device probably will only work on a wooden gate or door. Screw the metal trim right to the gate, placing one bare-ended hookup wire under each strip before it is tightened all the way. The electric doorlatch operates at 117v AC and is not too hard to purchase or install.

The location of the box doesn't matter a great deal, although the closer the box is to the gate, the less hookup wire is needed. Because it's weather-proofed, you can bury it completely, partially, or maybe even attach it to the fence. Camouflage the wires going to the metal trim and the doorlatch by stapling them to the gate wherever necessary and then painting them. You could also bury the cord to the wall socket to keep it out of the way.

Many other uses suggest themselves for this device. If you have an outdoor pool surrounded by a wooden fence, this could be ideal for summer use. A wooden door inside the house is not beyond the scope of this locking technique, either. Strips of foil or closely-spaced wires can be located along the top of the door. A very thin strip of metal would serve to close the circuit and open the lock. The same can hold true for any cabinet you want to keep locked.

22
Splash Alarm For Swimming Pool

When you're a kid, one of the highlights of the long, hot summer vacation is going swimming. This could mean splashing around in the local mud hole, or if the climate is always hot and backyard pools are a common convenience, it could mean some frolicsome fun in the neighbor's pool. Or, if you're fortunate enough to have your own backyard swimming pool you have no doubt played host to the neighborhood mob many a time. This is all fine and dandy, but one of the most urgent requirements, of course, is having responsible supervision present, capable of rescuing a child and taking the necessary measures to insure his safety. The splash alarm described here is ideal for letting you know if someone is in the pool at any time of the day or night. When a couple of neighborhood rascals decide to invade your pool to do a little midnight "skinny-dipping," you'll know.

The main parts in the project are an easily constructed float,

PARTS LIST

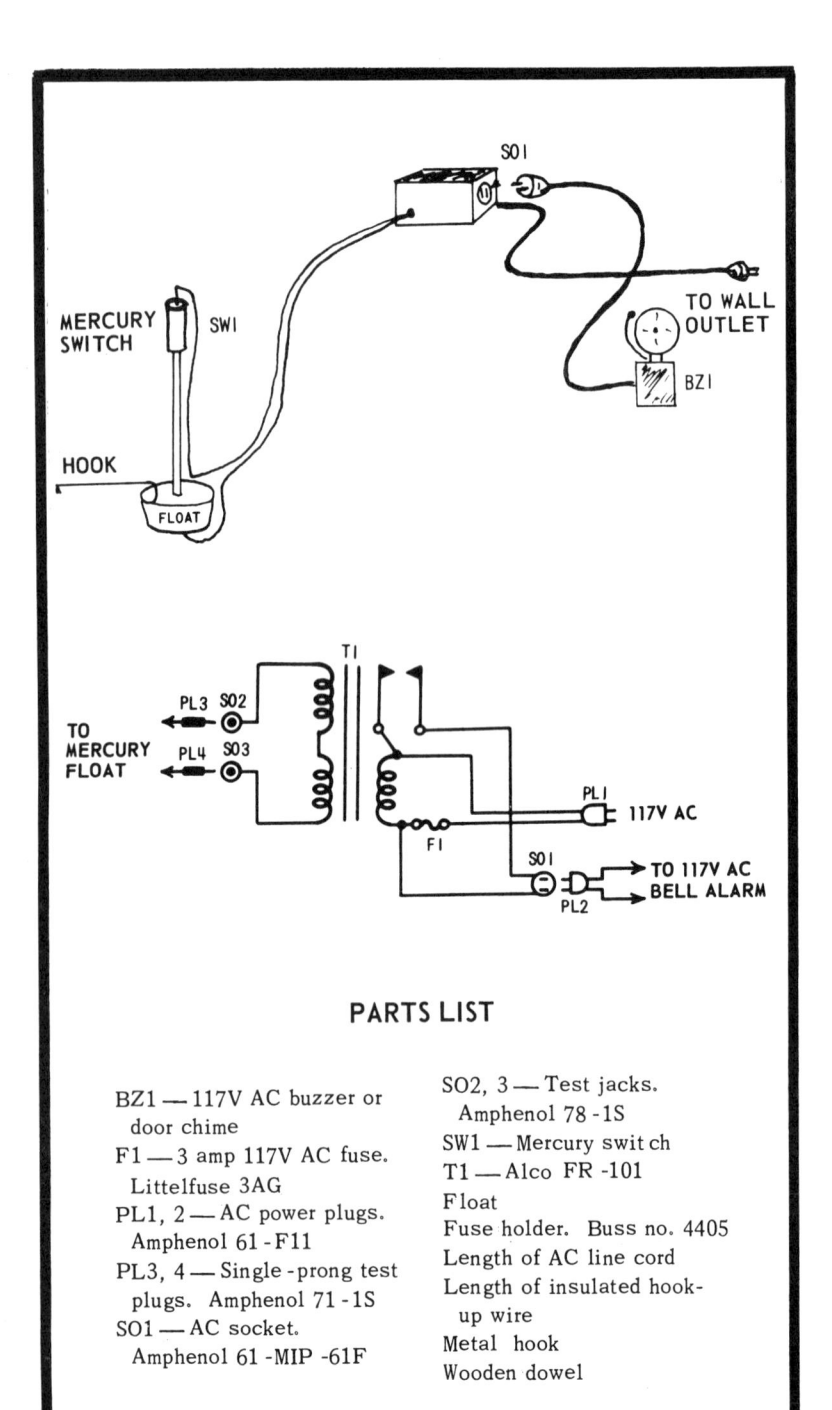

BZ1 — 117V AC buzzer or door chime
F1 — 3 amp 117V AC fuse. Littelfuse 3AG
PL1, 2 — AC power plugs. Amphenol 61-F11
PL3, 4 — Single-prong test plugs. Amphenol 71-1S
SO1 — AC socket. Amphenol 61-MIP-61F
SO2, 3 — Test jacks. Amphenol 78-1S
SW1 — Mercury switch
T1 — Alco FR-101
Float
Fuse holder. Buss no. 4405
Length of AC line cord
Length of insulated hook-up wire
Metal hook
Wooden dowel

a mercury switch, some kind of alarm device, and the isolation relay. The alarm operates very simply. When the water in the pool is disturbed, the float will rock back and forth, which in turn causes the mercury in the switch to flow back and forth (see pictorial). If the disturbance is great enough, the flowing mercury will short out the secondary, which activates the alarm. The connections are only momentary, the switch remaining open until the float rocks the other way. The design of the float will depend on the use that you have for it. If you make the bottom circular part larger, it will be less responsive to waves of a given size. Make it smaller and it will bob more. Different materials make a difference, too; e.g., styrofoam will bob more than wood. The thickness of the float will also have an effect. Thus, it will be necessary to experiment to get the best arrangement for the size of your pool. The longer the vertical dowel, the greater the angle of movement it will go through and the more sensitive it is. You have a limit though, because it does have to float. A small hook made out of coat hanger wire will keep the float near the edge of the pool.

You might want to build your float before you buy the mercury switch. They come in different sizes and consequently make some difference in the sensitivity of your alarm.

Mount the isolation relay, fuse, and socket in a small metal box. The mercury switch operates on low-voltage, so you can use weather-proofed insulated hookup wire to connect it. Be sure to cover the connections with electrician's tape to protect them from the water. For an alarm, use any 117-volt AC type. The chime that is used for a door bell is suggested because of its availability.

Now you're all ready to go. When you want to use the pool just remove the float from the water or unplug the alarm. Don't forget to connect it again, though! Also, if you have ever kept large tropical fish, you know that years of work can go down the drain if they get into a serious fight. Here is another use for your float. Make a miniature version that will tell you when your fish are creating a rukus, before it's too late.

Press-to-Talk Tape Recorder Switch 23

Nothing is more frustrating when you're recording than to have to leave the place where you're sitting every time you want to temporarily turn your recorder on or off, or to have your tapes interrupted by the noisy clicks that are caused by pushbutton operation. This simple and safe switching assembly enables you to have the controls of your recorder at your fingertips—across the room, in another room, or right next to the recorder. The small size of the switch eliminates any troublesome pushbutton clicking that you may have heard on your tapes. You can do away with a myriad of everyday inconveniences that you may not even have been aware of. The different switches that you can use, and the consequently different functions of the circuit that may be of use to you, will be discussed later.

The project consists mainly of a combined step-down transformer and relay mechanism (the Alco FR-101 is recommended) and a pushbutton switch. The primary coil, as you can see on the schematic diagram, is connected directly to 117 volts AC. The lines to the recorder will also receive this voltage, but only when the relay is closed. A constantly changing voltage in the primary coil induces a voltage in the secondary. Being a step-down transformer, the secondary voltage is much lower than the primary, having a maximum value of only 4 to 6 volts. This voltage, if the circuit is closed, will induce a current, causing the relay to snap shut. This completes the circuit to the recorder. If the secondary circuit is broken, the relay opens, too. It is important to note that the switch assembly regulates only the motor, not the whole

TO RECORDER MOTOR
NOT AC LINE

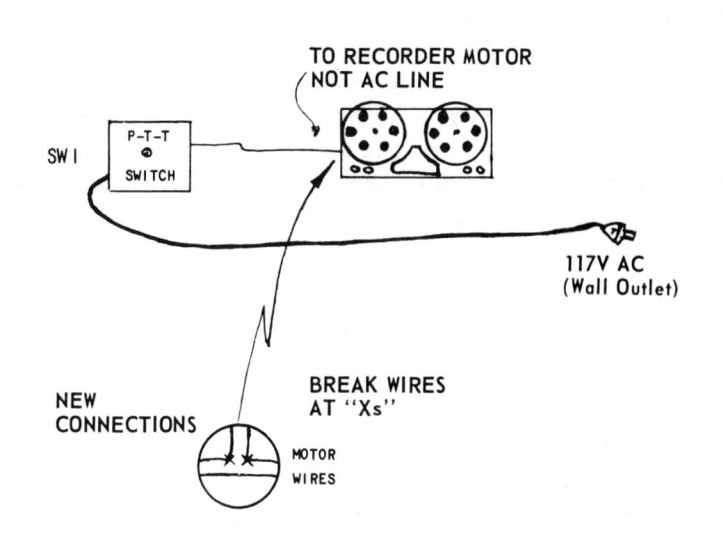

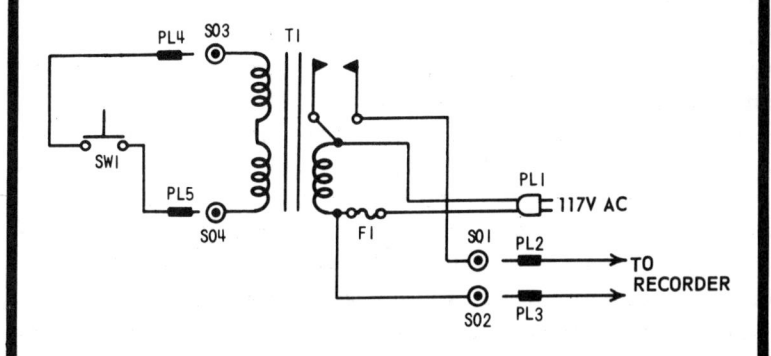

SW 1 — P-T-T SWITCH

117V AC (Wall Outlet)

NEW CONNECTIONS

BREAK WIRES AT "Xs"

MOTOR WIRES

PL4 SO3 T1
SW1
PL5 SO4
F1
PL1 — 117V AC
SO1 PL2 → TO
SO2 PL3 → RECORDER

PARTS LIST

F1 — 3 amp 117V AC fuse. Littelfuse 3AG
PL1 — AC power plug. Amphenol 61-F11
PL2, 3, 4 — Single-prong test plugs. Amphenol 71-1S
SO1, 2, 3, 4 — Test jacks. Amphenol 78-1S
SW1 — Normally-open SPST pushbutton switch (see text)
T1 — Alco FR-101
Fuse holder. Buss no. 4405
Length of AC line cord

recorder. When it is wired this way, the recorder remains connected and in an operating condition while the motor is temporarily disconnected.

You might ask why we bother with the transformer and the relay. Why not just put a switch in the recorder circuit? Well, the transformer-relay is advisable because of a factor that is too often neglected—safety. An ordinary switch would work, but it wouldn't be safe. With this device the only portion of the circuit that you come into contact with is the low-voltage secondary, a factor which tremendously reduces the hazard encountered with high voltage. A fuse is included in the circuit to add further to the safety factor. If you have children in the house, the possibility of them unknowingly endangering themselves is thus eliminated.

As was mentioned before, there is more than one type of switch that may be used in the secondary circuit. One kind is a normally-open SPST pushbutton switch. This will give you the press-to-talk feature for recording. Unless you press the button the secondary remains open and the recorder motor remains off. This is ideal for dictating short messages or taping single pieces of music. If you want to turn your recorder temporarily off instead of temporarily on, use a pushbutton switch that is normally-closed instead of open. Sit back in a comfortable chair, turn on your favorite radio program, put the switch on the arm of your chair and you are prepared to record to your heart's content. The commentator's voice can be conveniently eliminated by pressing the button at the termination of each piece of music.

Recording a speech or a discussion at home can be a problem if you have little kids running around the house. The tape can be ruined while you fumble for the button to turn the recorder off if you're interrupted. With this device, you can have the controls at your fingertips. If you want to pause briefly while recording to review notes, this is for you.

Still another type of switch that you could use would be just an ordinary SPST. Remaining either open or closed, this switch will perform the same job as either of the two pushbuttons mentioned before. It does not, however, spring back to the original position.

Mount the circuit components in a small metal box. The wires to the recorder should be connected as shown in the diagram, breaking only one of the wires to the motor. All of the wires in the circuit, including those inside the box, are AC line cord. The cords to the wall outlet and the recorder should be made as long as necessary. If you don't want the connections with the recorder to be permanent, phone jacks could be used with an off-on switch between them.

If you don't like how the box looks as it is, there are many ways to improve its appearance. Covering it with attractive paper is simple and cheap. A handsome, thin-walled wooden box is another idea. Painting it is possible also. As you can see, the possibilities go on and on. The box can be mobile, or if recording is always done in the same place you can attach it anyplace where it will be easily accessible and convenient for you to use. This might be under a table or even right on a chair. The innumerable other applications that you might have for this circuit are left to you.

24
Low-Voltage Thermostat for Attic Fan

During the summer the attic can be a very unpleasant place to be. Even when the temperatures outside aren't too bad, the attic—because of poor ventilation and the heating of the sun—is apt to be 20 to 30° hotter. This makes it almost impossible to go up there during the day. In many houses, if the attic had an adequate ventilation system, it could be just as useful as any other room in the house. This can be accomplished by a simple device that will keep a fan running in the attic only when certain preset critical temperatures are reached. This way, you don't have to be running the fan all the time and won't be wasting electricity. Other useful applications for the same circuit are an incubator temperature

PARTS LIST

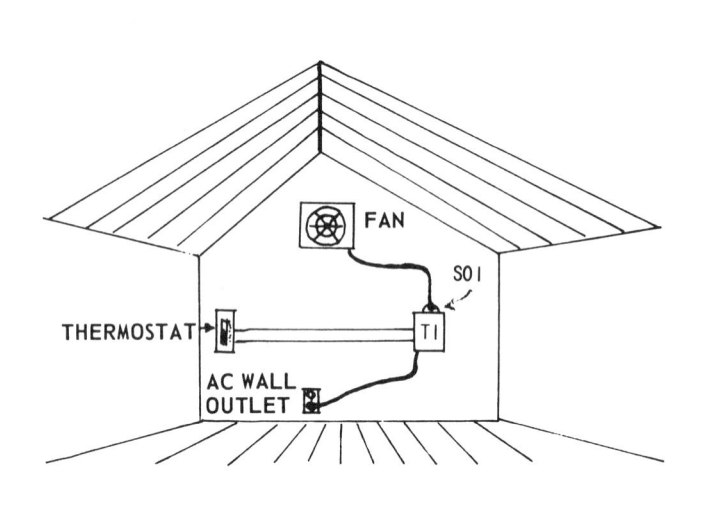

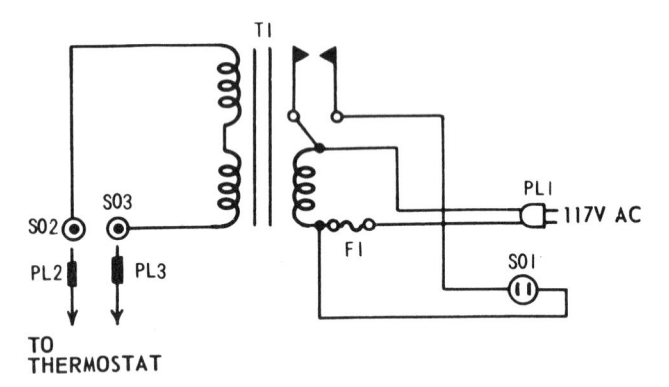

F1 — 3 amp 117V AC fuse. Littelfuse 3AG
PL1 — AC power plug. Amphenol 61-F11
PL2, 3 — Single-prong test plugs. Amphenol 71-1S
SO1 — AC socket. Amphenol 61-MIP-61F
SO2, 3 — Test jacks. Amphenol 78-1S
T1 — Alco FR-101
Fuse holder. Buss no. 4405
Length of AC line cord
Length of insulated hook-up wire
Thermostat, hardware store variety

control, aquarium light regulation, and temperature regulation in rooms other than the attic.

The principal components in the project are a low-voltage thermostat (hardware store variety) and a combined step-down transformer – relay mechanism (the Alco FR-101 is recommended). Since the transformer steps the voltage down, the constantly changing voltage in the primary coil induces a much lower voltage in the secondary coil. If the secondary circuit is closed, the resultant voltage produces a current. And whether it is closed or not depends solely on the thermostat which, in turn, depends on the temperature. If a critical temperature is reached, current will flow, causing the relay to snap shut—thus completing the circuit to the fan or whatever is plugged into the AC socket (SO1).

The reason for using the transformer-relay instead of putting the thermostat right in the high-voltage circuit is for safety considerations. In many cases, the thermostat might be placed at the opposite end of the room from the fan for maximum effectiveness. This means exposed wiring running the length of the room. As you can see, it is much safer to have a low voltage and current here. Also included in the circuit is a fuse to further safeguard against high voltage.

Mount the circuit components in a small metal box. All connections inside the box should be made with insulated hookup wire. The connection to the wall outlet, however, must be AC line cord. The wires to the thermostat can be ordinary bell wire because of the low voltage.

If you don't like how the box looks as it is, there are many ways to improve its appearance. You could cover it with attractive paper. A handsome wooden box is another idea; the possibilities are limitless. You can attach the box wherever you want, as long as it is near enough to the appliance it will serve.

The thermostat for the attic is the normally-open type. You can adjust it for the most desirable temperature and it will remain open at all temperatures below this critical one. If you have the fan in an attic window, you might want the thermostat at the other end of the room for best results. It is important, however, to keep it out of the sun if you want accurate response to the temperature in your attic.

If you keep tropical fish you know what a problem overheating of a tank from sunlight can be. In many cases this is caused by having the aquarium lights on at the same time that the sun is striking the tank. The problem can be solved by using a normally-closed thermostat instead of a normally-open one, as used in the fan circuit. Place the thermostat on the front of the aquarium and adjust it so that the overhead aquarium lights interract with the sunlight. When the sunlight strikes the tank, the thermostat will open and the tank lights will remain off until the sunlight disappears.

The same idea will work for an incubator. If you have ever been faced with the problem of maintaining desert animals in your home, you realize the necessity of keeping a high and constant temperature. The normally-closed thermostat and transformer-relay switch are ideal for this. An incandescent light bulb serves adequately as a heating element. When the temperature rises beyond a set value, the bulb is disconnected by the thermostat. If the temperature begins to fall again, the light will go back on.

Now you can have a cooler attic, a cooler aquarium, and a warmer terrarium. Actually, you can regulate any appliance in your home with this device—just plug in a radio, a clock, or a TV and it will work only at a certain temperature.

25
Mat Switch for 117v AC

Have you ever become impatient while waiting for someone to answer a doorbell. It may be that the individual you are calling on has to come all the way up from the basement or from the most distant room of the house? Maybe he didn't even hear the doorbell at all, being too far away or too engrossed in what he was doing. Here is a device that will make you more aware of the people approaching your house, so that

MAT SWITCH

S01

TO WALL SWITCH
(117V AC)

BZ1

PL1   S01   T1

SW1

MAT SWITCH   PL2
S02

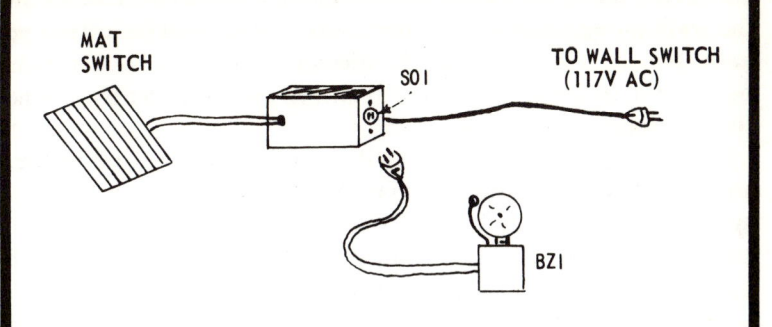

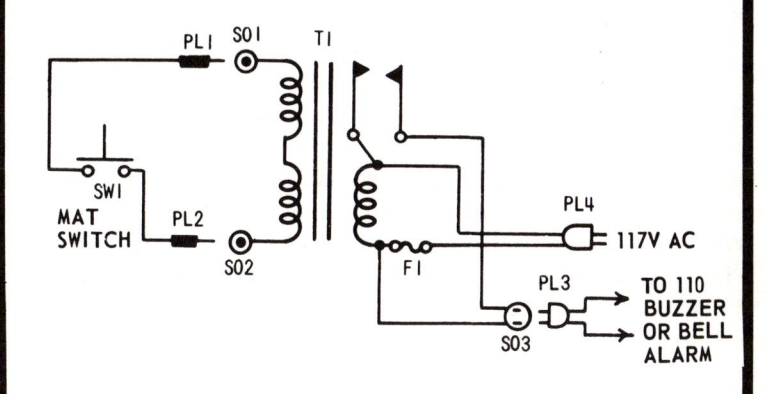

F1

PL4
117V AC

PL3
S03

TO 110
BUZZER
OR BELL
ALARM

PARTS LIST

BZ1 — 117V AC buzzer or alarm bell, hardware store variety
F1 — 3 amp 117V AC fuse. Littelfuse 3AG
PL1, 2 — Single-prong test plug. Amphenol 61-F11
PL3, 4 — AC wall plug. Amphenol 61-F11
S01, 2 — Test jacks. Amphenol 78-1S

S03 — AC socket. Amphenol 61-MIP-61F
SW1 — Mat switch. Record NF-28
T1 — Alco FR-101
Fuse holder. Buss no. 4405
Length of AC line cord
Length of insulated hook-up wire

you can be on your way to the door before they even ring the doorbell. When the caller steps on a mat placed part way down the walk going up to your house, a bell will sound, letting you know that he's coming. If, by chance, you don't have a doorbell, this is ideal for you. It might even let you know when the newspaper has come.

The main parts are a mat switch, a 110v AC bell or buzzer, and the isolation relay. Since the mat switch is normally-open, the secondary circuit will be complete only when the switch is depressed, for instance when someone steps on the mat that it's under. When this happens, the relay snaps shut and the bell or buzzer sounds. The switch will be stepped on only momentarily and then will snap open again, so you don't have to turn anything off.

You can mount the socket, isolation relay, and fuse inside a small metal box. That is, of course, because of the high voltage involved. The transformer steps down 117v AC to less than 30v AC. Because of this low voltage in the secondary, ordinary insulated hookup wire can be used. If you're going to use your switch where both it and the wires connecting it will be exposed to the weather, weather-proofed insulated wire might be a good idea. The connection from the box to the wall outlet is, of course, AC line cord.

Where you put your mat switch depends on the type of warning system you want. If you place it right on your doorstep it probably will let you know when your newspaper comes. If you place it part way down your front steps you'll catch the mailman, milkman, solicitors, and anyone else that comes up the walk. If you put it where nobody should step, under the bushes maybe, you'll find out if somebody does!

Remember, the whole idea is to be alerted when you are nowhere near the door. This would mean that you should probably place your warning bell down in the basement or in that distant room where the doorbell can't be heard. The wires to the mat switch can be as long as necessary. Running them in through a basement window is a good idea if that's where your alarm will be.

Many other ideas suggest themselves. You could place your mat switch near the entrance to a room. All you would have to do would be to walk in and the lights would go on. For this adaptation, a 2-mfd capacitor and a normally-closed pushbutton switch are added to the circuit. These can be wired exactly the same as in Projects 20 and 26. Garage lights could be turned on this way when the car enters and passes over the switch. The innumerable other applications are left to you.

26
Coin-Actuated Electric Switch

Very important in a child's education is his learning to make decisions and to distinguish between the relative importance of different activities and possessions. The allowance that he gets is generally only large enough to buy a soda or two at the local drugstore, to go and see an exciting movie, or perhaps to buy a new model airplane. He must, of course, decide which he would like most.

Here is a device that will be valuable to both your children and to you. With it you can make it necessary to pay a nickel or a dime to turn the TV on. Your son or daughter will have to decide whether it's worth it to watch a certain TV show. Not only will they learn to be more thoughtful when spending their meager assets, but they might be able to find more useful things to do with the same time and money. Even you can make use of this coin-actuated switch. As you know, nothing is easier than saving a little bit at a time. Paying for a TV by paying a little something every time you watch it is thrifty and easy on the pocketbook.

The project consists mainly of an easily constructed coin-switch, electrolytic capacitor, normally-closed pushbutton switch, and the isolation relay. Because of the capacitor in the secondary, AC current flows all the time. This alone is not enough to close the relay and to turn on the TV. The

PARTS LIST

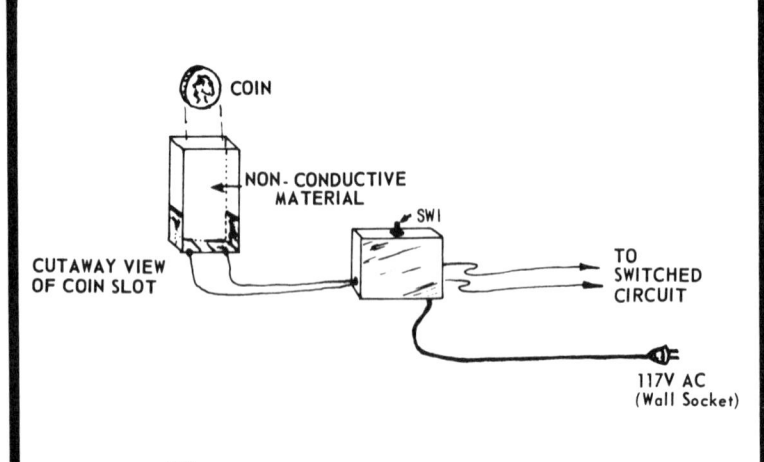

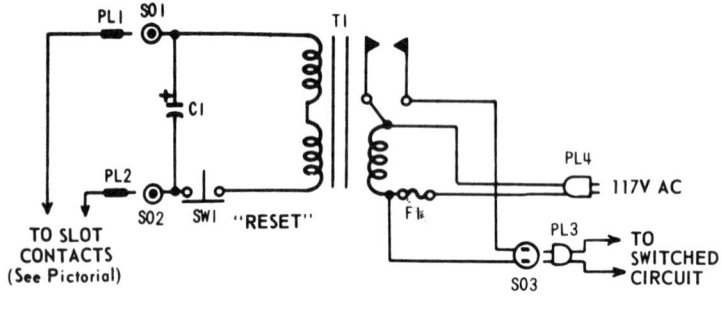

C1 — 2 mfd, 250 WVDC electrolytic
F1 — 3 amp 117V AC fuse. Littelfuse 3AG
PL1, 2 — Single-prong test plugs. Amphenol 71-1S
PL3, 4 — AC power plugs. Amphenol 61-F11
SO1, 2 — Test jacks. Amphenol 78-1S
SO3 — AC socket. Amphenol 61-MIP-61F
Fuse holder. Buss no. 4405
Length of AC line cord
Length of insulated hook-up wire
Stiff cardboard or phenolic, used to fashion coin slot
SPST pushbutton switch
SW1 — Normally-closed
T1 — Alco FR-101
2 Pieces flexible copper sheeting

capacitor inhibits the flow of current somewhat. When a coin closes the switch, the circuit is clear and enough current flows to close the relay. If the coin-switch is opened again, the current in the capacitor circuit is sufficient to keep the relay closed. In other words, the current across the capacitor is not great enough to close the relay, but it is strong enough to keep it closed. Pressing SW1 cuts off all current in the secondary and the relay opens.

If you do want to make the switch a bank, a coin "shute" can be connected to a metal or cardboard box. Make the shute two or three inches long and just large enough for the coin that you want to use. Make the two halves of the switch out of very narrow strips of flexible copper sheeting. The shute is positioned vertically so that a coin can roll, not slide, through it. Place the sheeting at the bottom of the shute end-to-end, but not touching. When a coin rolls down, the two pieces of copper will be momentarily connected, thus closing the relay. You can arrange it so that the nickel, or whatever, rolls right into the box. You'd better lock it or something to make sure you or the kids are not tempted to retrieve the funds after they're deposited. If you don't like the idea of saving, construct the coin-switch with a slot and the two electrodes as shown in the diagram.

Mount the isolation relay, fuse, and switch (SW1) in a small metal box. All high-voltage connections should be made inside this box. Use ordinary insulated hookup wire or bell wire for the connections to the shute. The voltage there is low and not at all dangerous. The connection to the wall is, of course, AC line cord.

Now you're prepared to avoid absolute hypnosis in front of the "boob tube." Both you and your kids will benefit from the funds deposited in your home-made bank. The other angle is to let your kids keep the money that they deposit in the TV. This is a great way to teach them to save their money.

27
Automatic Switch That Turns On Second Lamp

Here is an idea that you can have a little fun with. If a person isn't aware of your trick, he will get the impression that whoever did the wiring in your house was an absolute idiot. Whenever a particular light is turned on, anything that is plugged into the photocell-controlled slave unit also goes on—whether it be a light, radio, or electric can opener.

Equivalent to an ordinary cell in parallel with a resistor, a resistive photocell is the control mechanism in the secondary. Different resistance values will provide various sensitivities for the control unit. The relay, across which is connected the appliance or lamp (see pictorial and schematic), is closed when light of sufficient intensity falls on the cell.

Lightweight insulated hookup wire is good for the low-voltage secondary leads to the cell, which are safely left exposed. The base of the controlled lamp is an excellent location for the cell, although any place that is illuminated by the lamp should be all right—on a wall, under a piece of furniture, or even on your clothes! The place you choose ideally should facilitate concealment of the leads to the photocell, though.

To house the isolation relay and fuse a small box, preferably metal is suitable. Both the controlled light and the relay circuit must be plugged into the wall, while the slave appliance is plugged into a socket mounted in the side of the box.

While wires and cords are easily hidden under rugs, behind furniture, etc., concealing the box is apt to be a problem if you're trying to pull the wool over someone's eyes. You will have to experiment to discover the best arrangement for the room you have in mind. If you just can't seem to get results, try putting a magnifying lense over the photocell to

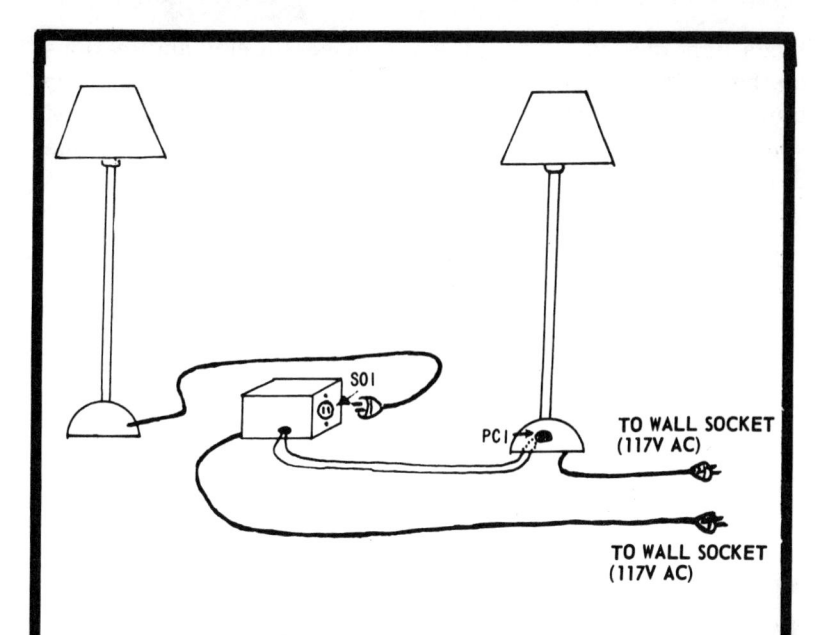

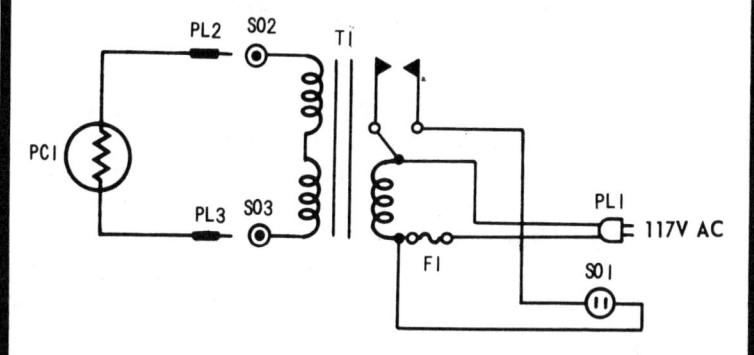

PARTS LIST

F1 — 3 amp 117V AC fuse. Littelfuse 3AG
PC1 — Resistive photocell. Lafayette 19 H2101
PL1 — AC power plug. Amphenol 61-F11
PL2, 3 — Single-prong test jacks. Amphenol 71-1S
SO1 — AC socket. Amphenol 61-MIP-61F
SO2, 3 — Test jacks. Amphenol 78-1S
T1 — FR-101
Fuse holder. Buss no. 4405
Length of AC cord
Length of insulated hook-up wire

focus light onto it. This is one of those "liquid" applications of the isolation relay. You will probably want to set it up temporarily and then go on to another project that has caught your attention. Versatility is a distinct advantage with this device.

28
Foolproof Smoke Alarm

Here is an ingenious way to protect your home from fire. Based on the premise that where there is fire there is also smoke, the alarm will respond not only to the billowing clouds of smoke from a kitchen grease fire but also to hotter fires that generate less smoke.

The heart of the project is a resistive photocell, which differs from the ordinary cell only in that it already represents a partially closed path, not an open vacuum condition. Therefore, less light falling on the cell will provide the necessary current to close the relay in the primary of the isolation relay.

As shown in the pictorial, the photocell is mounted in the end of a small tube, which can be a toilet paper tube lined with aluminum foil for optimum reflection. In the other end of the tube is mounted a magnifying lense, which is used to focus light onto the cell. Since the focal length of lenses varies, some adjustment for the best distance from the photocell should be made. The tube leads into a box, preferably wooden, which is painted a dull black on the inside to absorb rather than reflect light. On some kind of tray at the bottom the light socket is mounted. A sizable space should be pro-

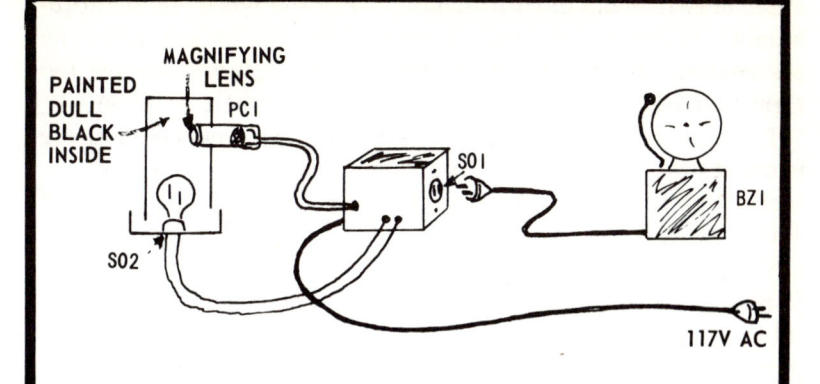

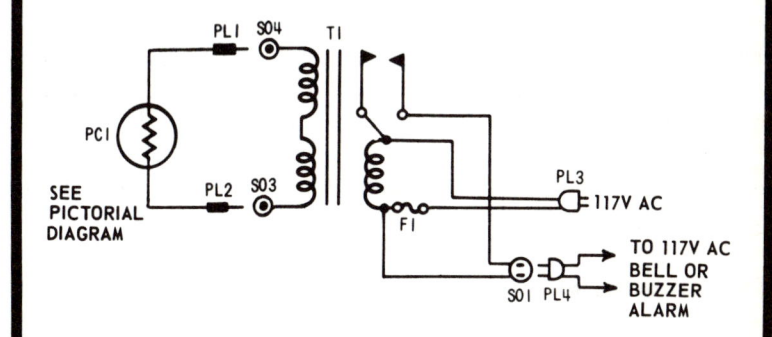

PARTS LIST

BZ1 — 117V AC buzzer or bell alarm, hardware-store variety
F1 — 3 amp 117V AC fuse. Littelfuse 3AG
I1 — Household bulb, low-wattage (15 watts or less)
PC1 — Resistive-type photocell
PL1, 2 — Single-prong test plug. Amphenol 71-1S
PL3, 4 — AC power plug. Amphenol 61-F11
SO3, 4 — Test jacks. Amphenol 78-1S
SO1 — AC socket. Amphenol 6-MIP-61F
SO2 — Porcelain light socket
SO3, 4 — Test jacks. Amphenol 78-1S
T1 — Alco FR-101
Fuse holder. Buss no. 4405
Homemade chamber, painted inside with dull paint. Notice the air gap between the base shield and the top chamber.
Length of AC line cord
Length of insulated hook-up wire
Magnifying lens

vided here to make certain that the air in the box is an accurate sample of the air outside. Circulation will be sustained by the heating effect of the bulb. Choose the best wattage for the light bulb after the project is complete (15 watts or less).

If a source of light were hidden from view, you would never know it was there unless you saw it reflected off some object. Such is the principle of operation of the set-up described above. Little light is reflected from the walls; consequently, any light entering the photocell tube will be reflected from particles in the air. If smoke density, and consequent light intensity, is great enough, the isolation relay will be triggered and the alarm will go off. As soon as the smoke clears up, the alarm will go off, due to the reduction in reflected light intensity.

Install the isolation relay and fuse in a small metal box. The socket for the buzzer or bell should be mounted in the side of the box. The light socket connections can go either to the box or directly to a wall outlet. Insulated hookup wire is fine for the exposed photocell wiring, since the secondary is low voltage. All exposed 117v AC leads should be AC line cord or some equivalent.

Experimenting with different arrangements for the tube, lense location, and light bulb wattage will give you the results you desire. Each different set-up will give your alarm more or less sensitivity to smoke particles.

The kitchen is probably the most valuable place for the use of this alarm. Because it is not practical to keep a light on all the time, using it only while cooking to avoid kitchen fires is not a bad idea. This, of course, is only one of the numerous creative applications.

PROJECTS FOR THE CAR

29

Electronic Tach For $5

Here's a handy device which will enable you to effectively improve the efficiency of your automobile. If your car has been slow to accelerate, your engine's RPM is probably not at its most efficient level. Solve your problem by accurately adjusting the RPM with this inexpensive instrument.

The project consists mainly of two transistors, a milliammeter, and a potentiometer. The scale on the milliammeter

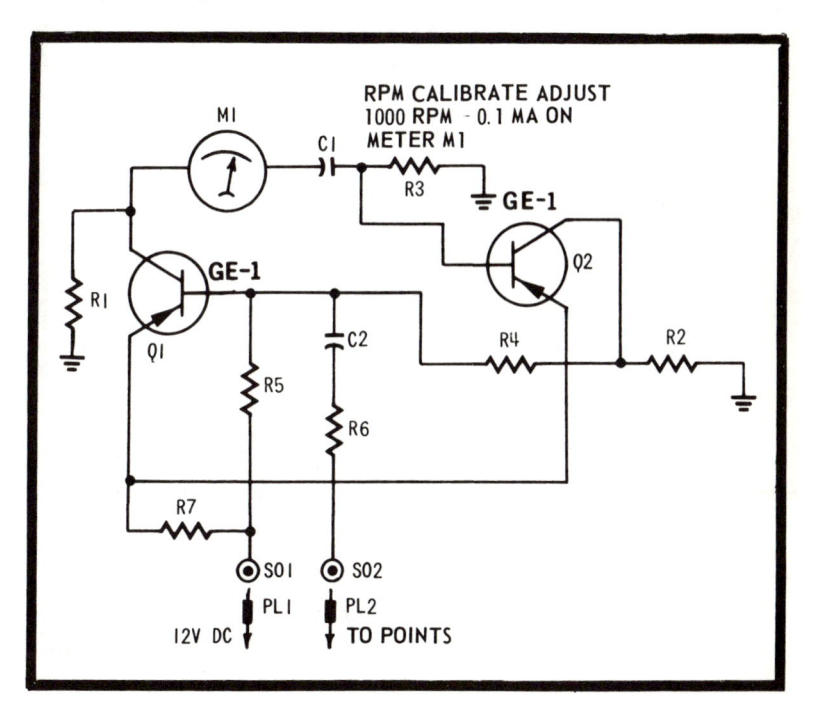

AIR HEAT VENT COOL

AUTO DASHBOARD

TACHOMETER
RPM

LIP FASTENED TO UNDERSIDE OF DASHBOARD WITH SELF TAPPING SCREWS

L- SHAPED LIP

HEAVY- GAUGE METER BRACKET

0 - 1 MA METER

CIRCUITRY

COVER (Can be Plastic)

RPM CALIBRATE ADJUST

BAKELITE ASSEMBLY BOARD

PARTS LIST

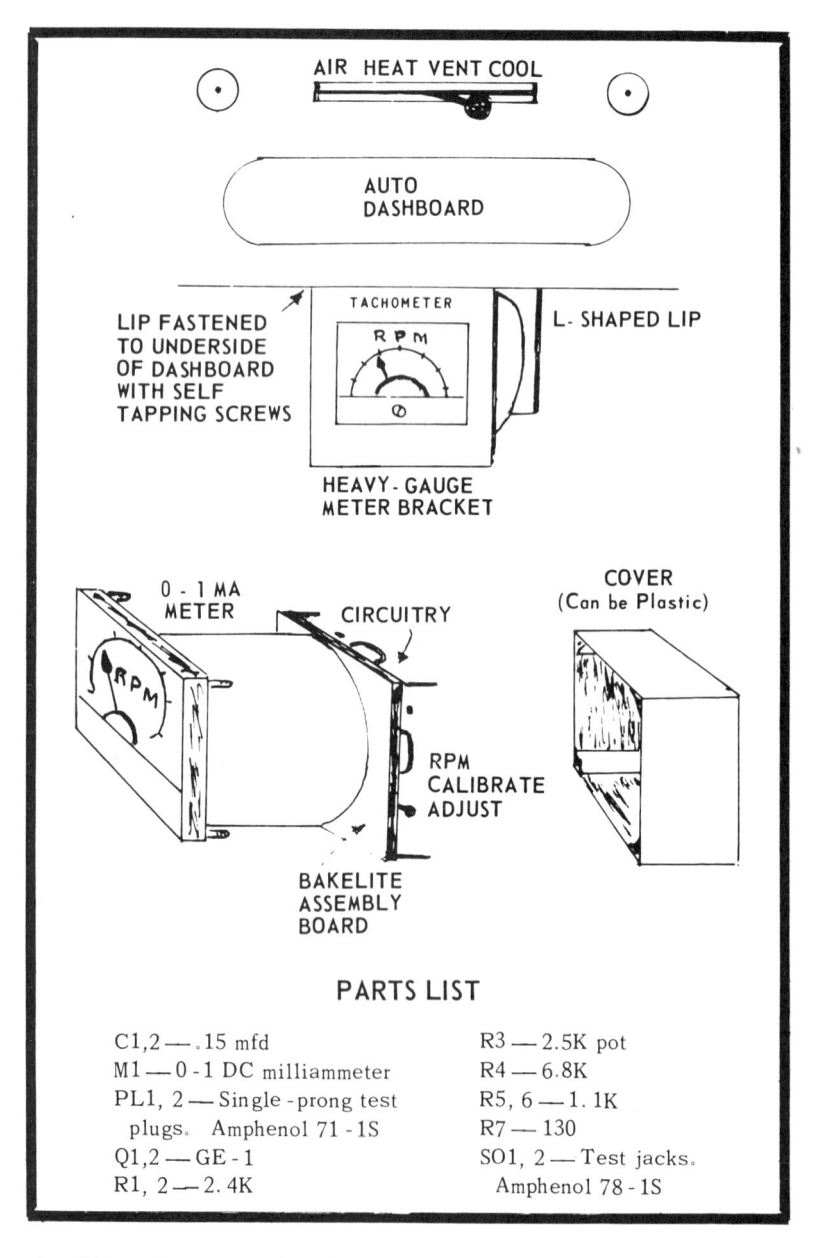

C1,2 — .15 mfd
M1 — 0-1 DC milliammeter
PL1, 2 — Single-prong test plugs. Amphenol 71-1S
Q1,2 — GE-1
R1, 2 — 2.4K

R3 — 2.5K pot
R4 — 6.8K
R5, 6 — 1.1K
R7 — 130
SO1, 2 — Test jacks.
  Amphenol 78-1S

should go from 0 to 1. The potentiometer (R3) is used to calibrate the instrument.

Be sure to calibrate your instrument according to the diagram. R3 should be adjusted so that M1 reads .1 ma at 1000 RPM.

30
Relay-Type Headlight Alarm

How many times have you experienced the misery and despair of not being able to get your car started because you forgot and left the headlights on? Too many, right? Here is a handy circuit that will safeguard you from this embarassing situation. If either the parking or the headlights are left on after the ignition is turned off, the relay is activated and the buzzer goes on, reminding you to turn off your lights before you leave the car.

The project consists of four diodes and a 5K SPST relay; the Lafayette 99R6091 relay works very well in this circuit.

You can construct the alarm in a Minibox, leaving the alarm

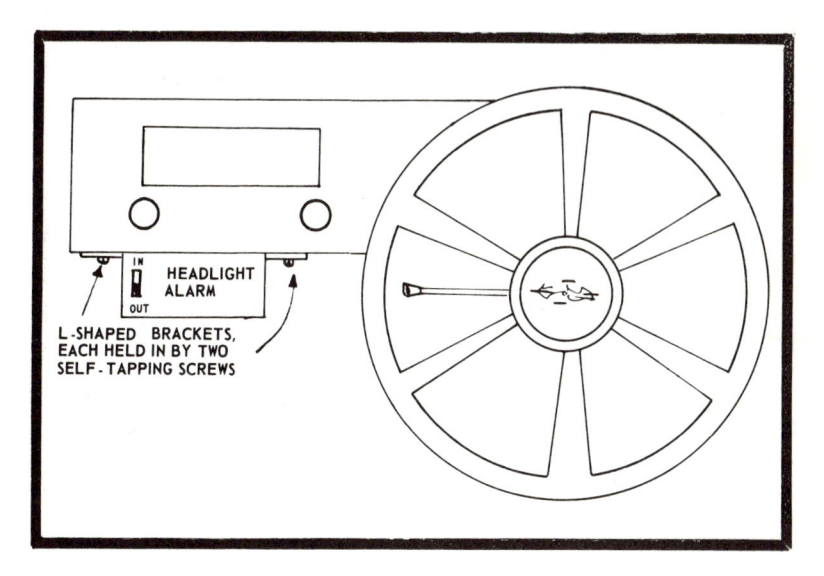

117

118

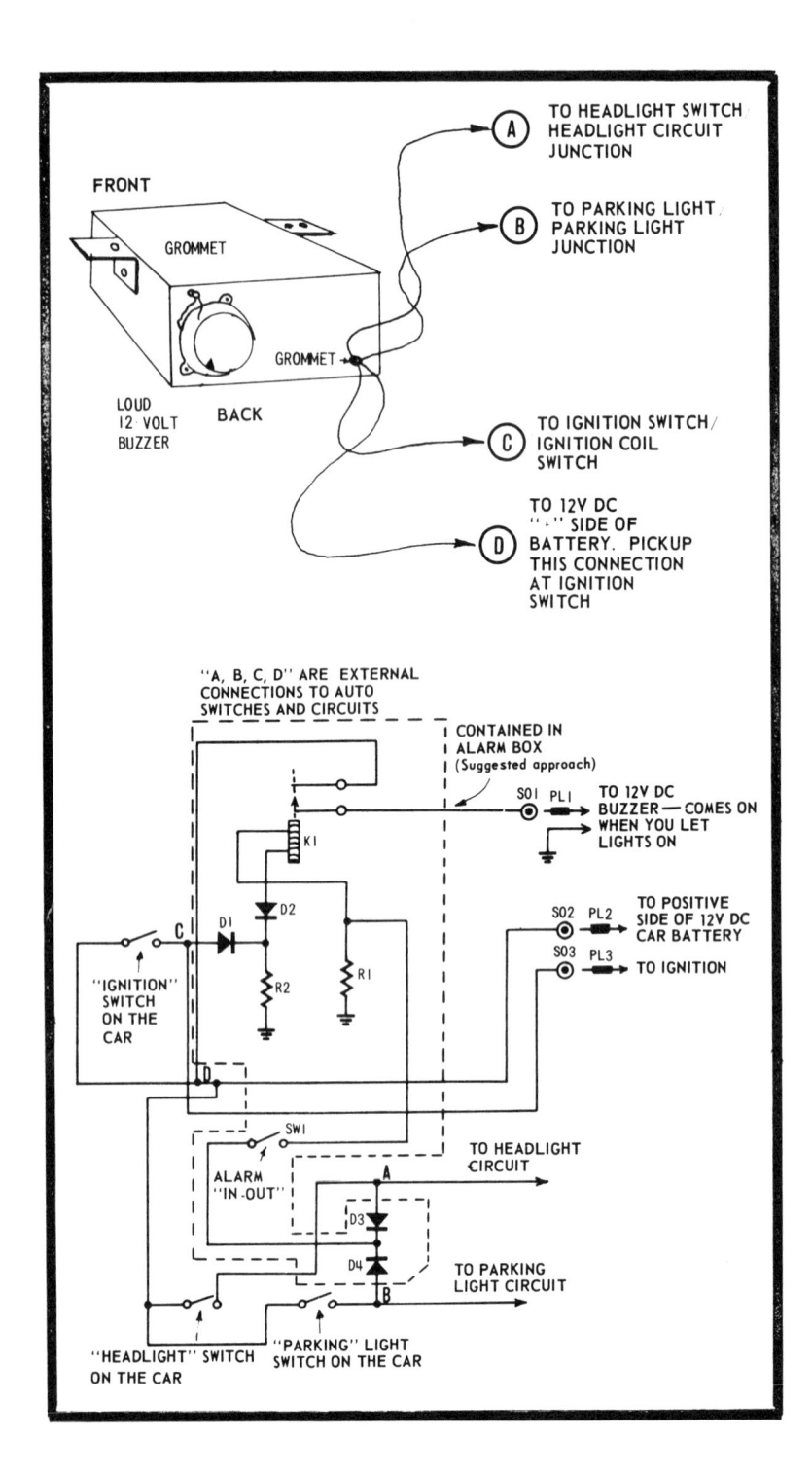

PARTS LIST

D1, 2, 3, 4 — 1N38B
K1 — 5K SPST relay.
   Lafayette 99R6091
PL1, 2, 3 — Single-prong test
   plugs. Amphenol 71-1S

R1, 2 — 2.4K
SO1, 2, 3 — Test jacks.
   Amphenol 78-1S
SW1 — SPST. Oak 200

---

switch (SW2) in a convenient place for turning the alarm on and off. Remember that this circuit is not the whole alarm, only the triggering device. Connect an inexpensive buzzer as shown in the diagram.

If, for some reason, you want to have your lights on while the engine is off, all you have to do is turn the alarm off. But, don't forget to turn it on again!

31 Wake-Up Alarm For Sleepy Drivers

Here is a handy gadget that will keep you awake while you're driving. At night, especially after many long hours of driving toward a vacation destination or business appointment, the extreme danger of falling asleep behind the wheel is ever present. It is hard to realize that you're falling asleep until it is too late, both for you and anybody who is unlucky enough to be on the road with you.

This advice operates so that if you become groggy and your head tilts in any direction, a tone will sound in the earphone, warning you before it's too late that you are on the verge of falling asleep.

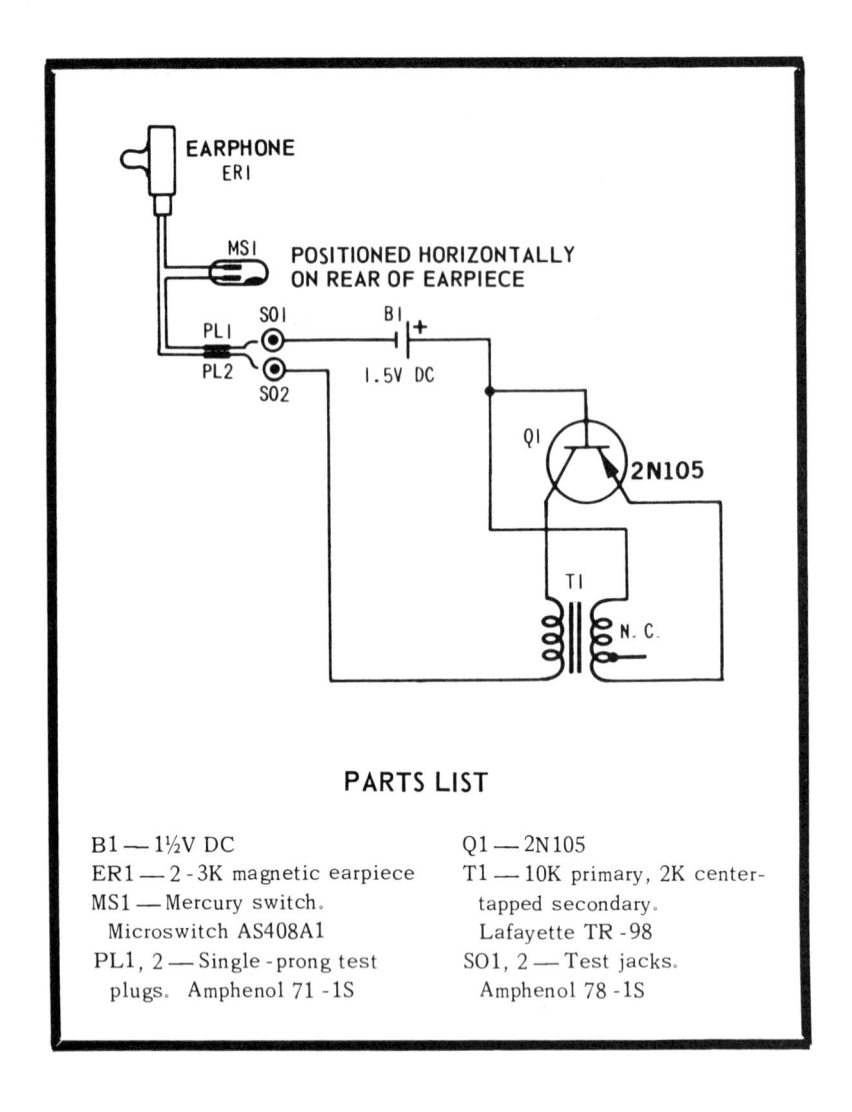

PARTS LIST

B1 — 1½V DC
ER1 — 2-3K magnetic earpiece
MS1 — Mercury switch.
   Microswitch AS408A1
PL1, 2 — Single-prong test
   plugs. Amphenol 71-1S

Q1 — 2N105
T1 — 10K primary, 2K center-
   tapped secondary.
   Lafayette TR-98
SO1, 2 — Test jacks.
   Amphenol 78-1S

The project consists of a transistor, a magnetic earpiece (2-3K), a mercury switch, and a transformer. The transformer, a Lafayette TR-98, has a 10K primary and a 2K secondary.

The heart of the project is the mercury switch, a microswitch AS408A1. Place it horizontally on the rear of the earpiece. The circuit, while normally open, is closed by the flowing mercury inside the switch when you tilt your head, causing the tone to be produced in the earpiece.

The circuit is battery powered, requiring only 1.5 volts DC.

You can assemble it in a small plastic box that can be conveniently placed in a shirt or coat pocket while it is in use.

32

Fox Hunt Transmitter Sniffer

Here is the simplest transmitter detector around. Made of only a few home-brew and easy-to-come-by components, the sniffer is suitable for any medium-range hidden radio transmitter ("fox") hunt.

C1 is the tuning capacitor and must be selected with a certain frequency in mind. Because the circuit is so very simple, you have to replace and substitute components to tune it. C1 must resonate with L1 and L2 on the transmitter's frequency.

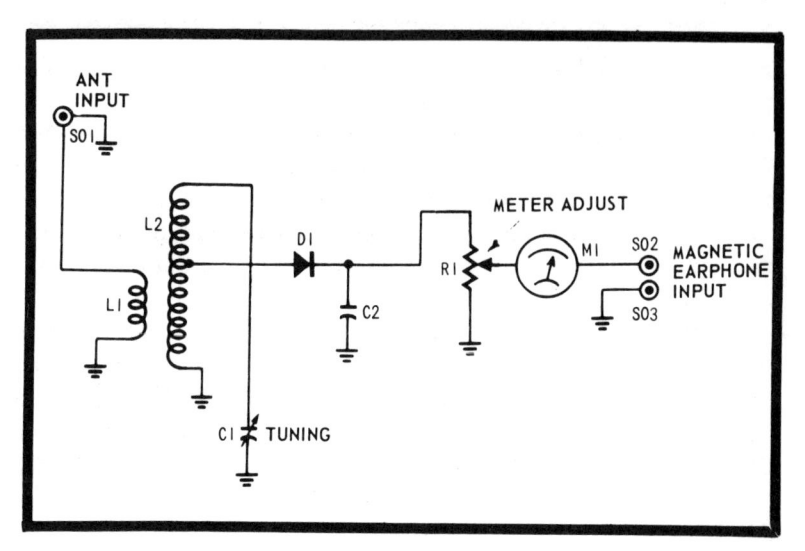

ANTENNA CONNECTOR

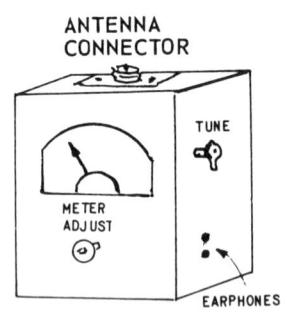

TUNE

METER ADJUST

EARPHONES

PARTS LIST

C1 — Select variable capacitor to resonate with L1 / L2 on the transmitter's frequency. Check with grid-dip meter. For 144-MHz operation, for example, you could employ a 2-plate VHF variable type with a capacity of approximately 0-15 pfd.
C2 — 100 pfd
D1 — 1N38B
L1 — 6-turn link, closely coupled with L2

L2 — Select to resonate with C1 on the transmitter's frequency. Check with grid-dip meter. For 144-MHz operation, you could use 4 turns of no. 16 enameled wire, 3/8" diameter, tapped about 1/3rd way up from grounded end.
M1 — 0-250 DC microammeter
R1 — 35K
SO1 — Coax receptacle. Amphenol 83-1R
SO2, 3 — Test jacks. Amphenol 78-1S

(See the parts list.) Check this resonance on the exact frequency with a grid-dip meter.

L2 also must be picked for the frequency of the transmitter that you're trying to detect and to resonate with C1. (See the parts list.) L1 is fixed—a 6-turn link coupled with L2. This coil combination should be checked with a grid-dip to insure the correct frequency. For frequencies other than 144 MHz, you will have to improvise.

R1, the sensitivity adjust, will put the signal that you pick up on the microammeter scale. The rectifier in the circuit reduces the minute incoming signal to the pulsating DC read on the meter. Assuming that your "fox" is modulated, there will also be an audible indication of directional signal strength in the magnetic earphone. A good directional antenna will greatly aid your sleuthing, too.

33

Idle Speed Calibrator & Tach

This device enables you to accurately calibrate the speed at which your car is idling. So what? The correct idling speed is of immense value to the operating efficiency of your automobile, especially in city driving where the car is idling more because of stop lights and traffic jams. This handy instrument will tell you, by showing only one of its two lights, that you have hit the correct idling speed.

The main parts in the project are a transformer, two bulbs,

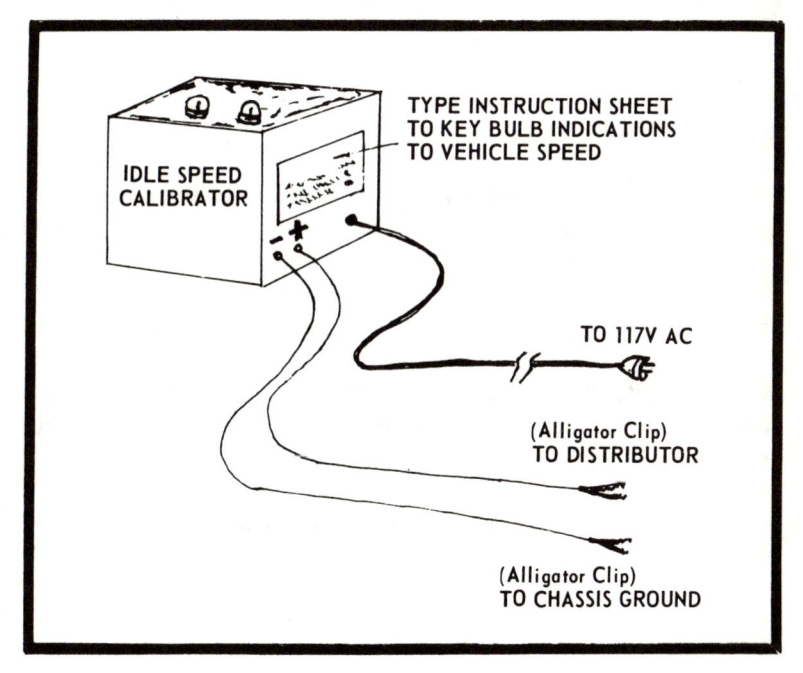

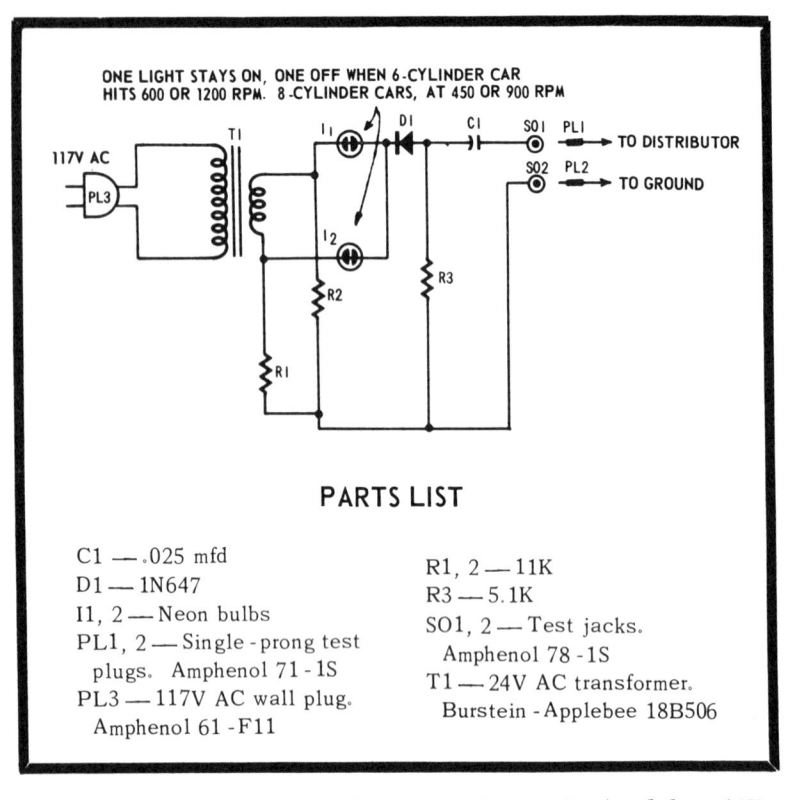

ONE LIGHT STAYS ON, ONE OFF WHEN 6-CYLINDER CAR HITS 600 OR 1200 RPM. 8-CYLINDER CARS, AT 450 OR 900 RPM

PARTS LIST

C1 — .025 mfd
D1 — 1N647
I1, 2 — Neon bulbs
PL1, 2 — Single-prong test plugs. Amphenol 71-1S
PL3 — 117V AC wall plug. Amphenol 61-F11

R1, 2 — 11K
R3 — 5.1K
SO1, 2 — Test jacks. Amphenol 78-1S
T1 — 24V AC transformer. Burstein-Applebee 18B506

and a rectifier. The transformer, a Burstein-Applebee 18B-506, steps down 117 volts AC to 24 volts AC, a usable voltage for this circuit. The indicator lights (I1, I2) are NE-51 neon bulbs.

Your calibrator will work differently for engines with different numbers of cylinders. When a 6-cylinder car hits 600 or 1200 RPM, one bulb on the instrument will be on and the other off. If you're checking an 8-cylinder car, one light will stay on and the other off when the engine hits 450 or 900 RPM.

As you can see, this instrument is very easy to operate. It is important, however, to use the specified values for the parts in the circuit in order to obtain accurate measurements and adjustments.

34
Six-Transistor Car Reverb

If you want to improve the sound of your car radio, here is the gadget for you: a rear speaker supplied with a 6-transistor reverb unit. With this, you can turn on your car radio and listen to the stereo effect and the rich, deep sound provided by the reverb unit. Once you've installed this unit, you'll be surprised at how much more fun driving has become.

The parts required for this project are a reverb unit (a Gibbs Type 5G is recommended), three TR-19 transistors (the International Rectifier Corp.), one 2N3706 transistor, two GE-3 transistors, a 3.2-ohm PM speaker, and various resistors and capacitors.

The circuit can be assembled in a metal Minibox and bolted to the bottom of the dash so that the knobs are easily accessible from the driver's seat. The Minibox can be painted to match the car's interior, if you wish.

It is very important that the polarities of the electrolytic capacitors are matched to the schematic and that the cables are shielded, so that no interference from the ignition occurs.

Because the car will be subjected to many bumps and constant vibration, the cables leading to and from the Minibox should either be protected by rubber grommets so that the insulation is not frayed against the sharp edges of the Minibox, or jacks should be placed in the holes. Also, the reverb unit should be shock-mounted so that it is not damaged.

Fader adjust, R18, determines the balance between the front and rear speakers, and the level control knob, R4, varies the amount of reverberation. Switches SW1A and SW1B are to be ganged. These switch the reverb unit in the rear speaker

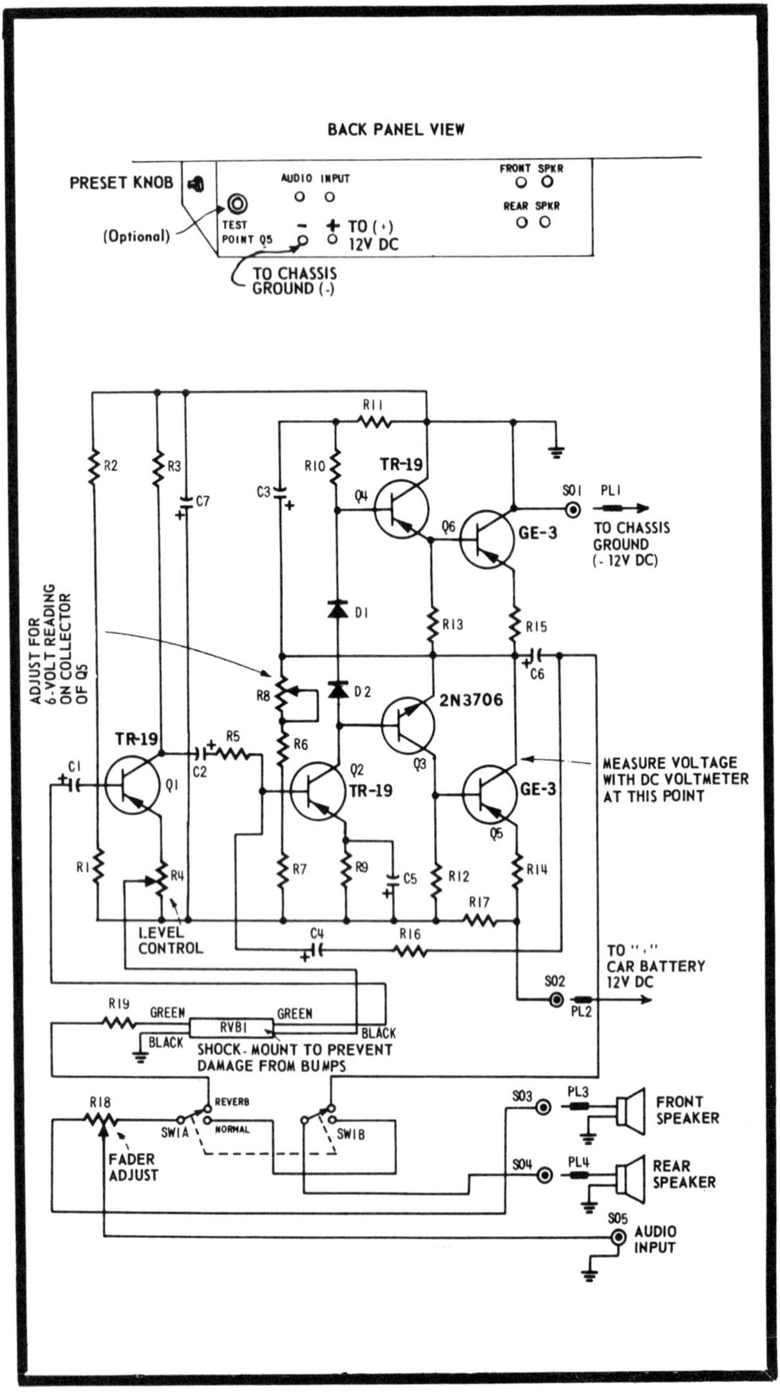
BACK PANEL VIEW
PRESET KNOB
(Optional)
AUDIO INPUT
TEST POINT Q5
- + TO (+) 12V DC
TO CHASSIS GROUND (-)
FRONT SPKR
REAR SPKR
ADJUST FOR 6-VOLT READING ON COLLECTOR OF Q5
TR-19
GE-3
2N3706
TO CHASSIS GROUND (- 12V DC)
MEASURE VOLTAGE WITH DC VOLTMETER AT THIS POINT
LEVEL CONTROL
TO "+" CAR BATTERY 12V DC
GREEN
BLACK
RVB1
SHOCK-MOUNT TO PREVENT DAMAGE FROM BUMPS
REVERB
NORMAL
FADER ADJUST
FRONT SPEAKER
REAR SPEAKER
AUDIO INPUT

VEHICLE DASHBOARD

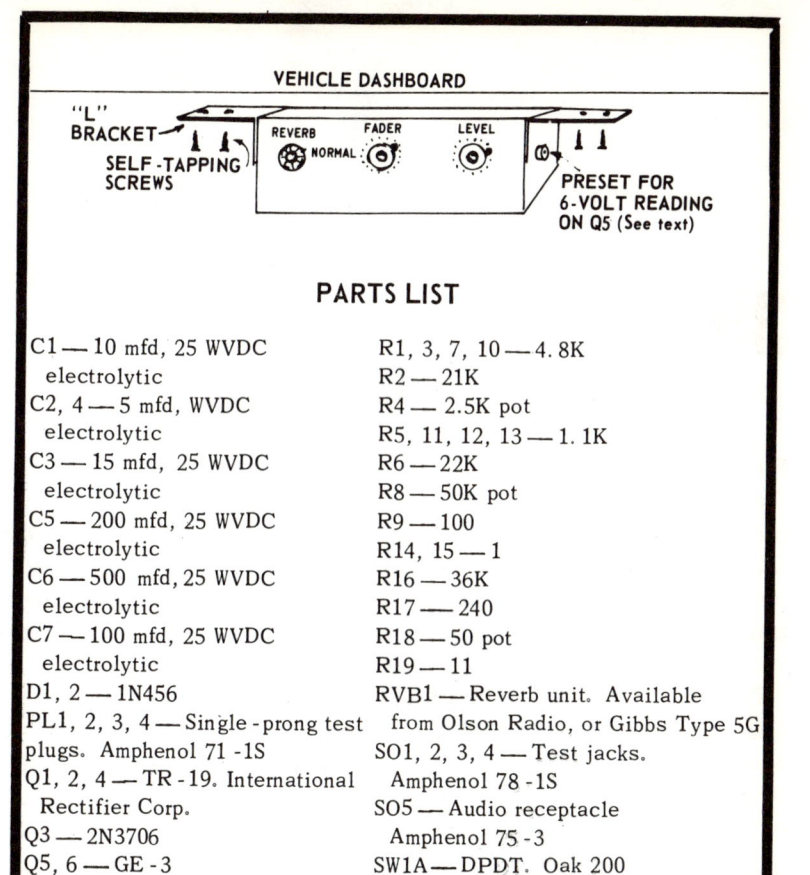

```
"L"
BRACKET        REVERB   FADER   LEVEL
SELF-TAPPING      NORMAL
SCREWS                                  PRESET FOR
                                        6-VOLT READING
                                        ON Q5 (See text)
```

PARTS LIST

C1 — 10 mfd, 25 WVDC electrolytic	R1, 3, 7, 10 — 4.8K
C2, 4 — 5 mfd, WVDC electrolytic	R2 — 21K
	R4 — 2.5K pot
C3 — 15 mfd, 25 WVDC electrolytic	R5, 11, 12, 13 — 1.1K
	R6 — 22K
C5 — 200 mfd, 25 WVDC electrolytic	R8 — 50K pot
	R9 — 100
	R14, 15 — 1
C6 — 500 mfd, 25 WVDC electrolytic	R16 — 36K
	R17 — 240
C7 — 100 mfd, 25 WVDC electrolytic	R18 — 50 pot
	R19 — 11
D1, 2 — 1N456	RVB1 — Reverb unit. Available from Olson Radio, or Gibbs Type 5G
PL1, 2, 3, 4 — Single-prong test plugs. Amphenol 71-1S	SO1, 2, 3, 4 — Test jacks. Amphenol 78-1S
Q1, 2, 4 — TR-19. International Rectifier Corp.	SO5 — Audio receptacle Amphenol 75-3
Q3 — 2N3706	
Q5, 6 — GE-3	SW1A — DPDT. Oak 200

on and off. Resistor R8 is to be preset for a 6-volt reading on the collector of transistor Q5 and is not to be an external knob. Except for these controls, the radio is operated normally from the present radio. Once installed, the reverb unit can be switched on whenever you please, providing you with a fine reverberation effect and constant enjoyment.

35
Universal Safety Flasher

Have you ever been in a hazardous situation where you wished that you had something more than just your car lights to let other drivers know that you were there? If you have, then this flasher is what you need. At night, if you are forced to change a tire on the edge of the road, or are carrying a protruding load in the back of your car, this flashing light will at least let other drivers know that they should be careful.

The project consists mainly of two transistors, two electrolytic capacitors and a 12-volt DC light bulb. The transistors and capacitors interact in a cycle of charging and discharging to cause the light to flash on and off. Be sure to connect the capacitors (C1, C2) as shown in the diagram.

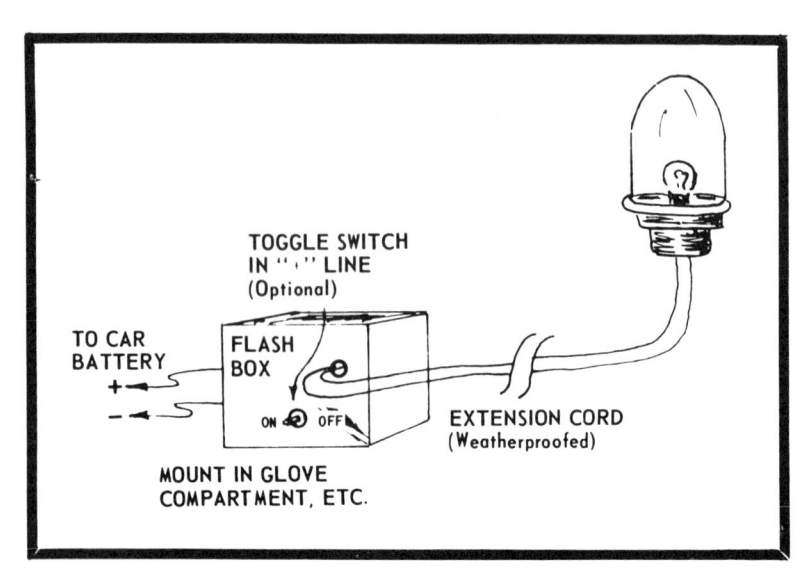

PARTS LIST

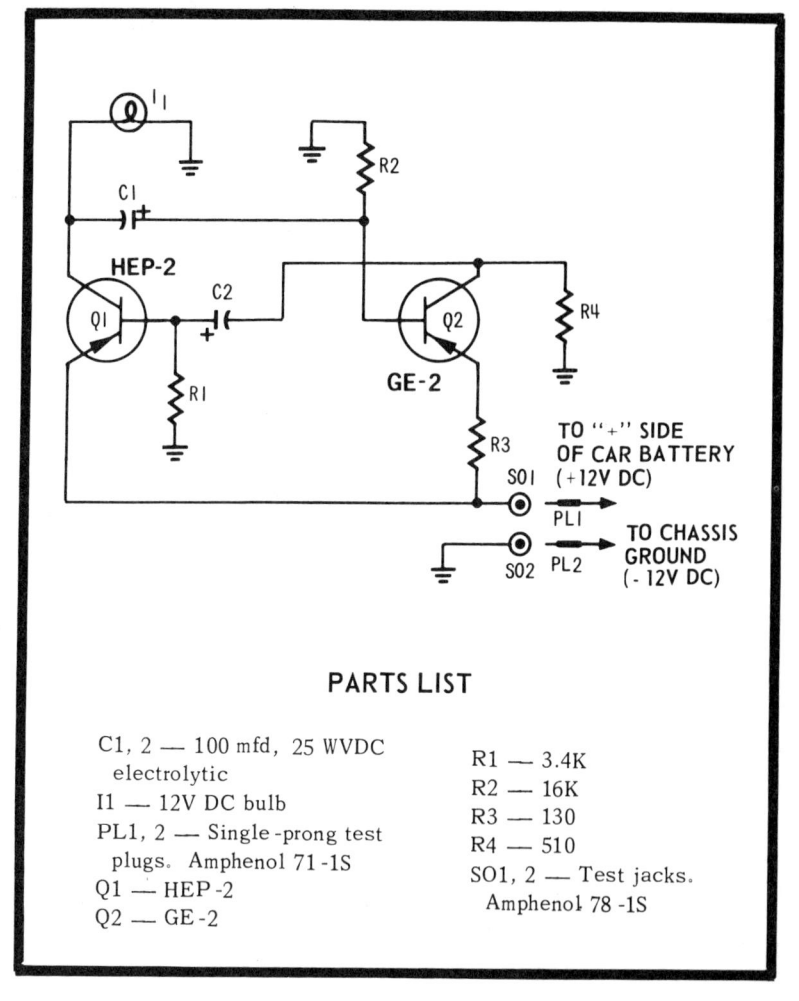

C1, 2 — 100 mfd, 25 WVDC electrolytic
I1 — 12V DC bulb
PL1, 2 — Single-prong test plugs. Amphenol 71-1S
Q1 — HEP-2
Q2 — GE-2

R1 — 3.4K
R2 — 16K
R3 — 130
R4 — 510
SO1, 2 — Test jacks. Amphenol 78-1S

You could mount the circuit in the glove compartment with the light bulb on a long extension wire. That way, you can place it wherever it is most effective for each set of circumstances. A transparent red plastic fixture for the bulb will facilitate its being seen.

129

36
Vibrator "Substitutor"

Here we have an interesting innovation that will do electrically the same thing that your old vibrator did mechanically. What difference does it make? You can eliminate the audible hum that it produces and the troublesome RF interference and hash that you may have heard on your car radio.

The main parts in the project are a modified transformer and two transistors. Unwind and connect the transformer according to the instructions on the diagram. Remember that for 6-volt Volkswagens only 60 turns of wire are removed and HEP-232 transistors are used instead of the GE-3 type. The circuit should be mounted in a metal container, such as a Minibox.

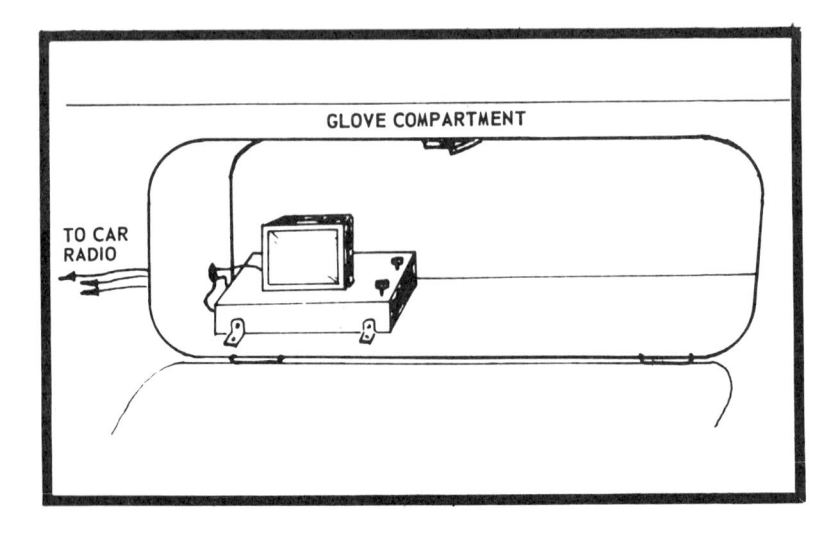

```
                    T1      YELLOW        GE-3
              RED                              Q1           TO VIBRATOR
                                                            SOCKET PIN 1
                              R1      R2                    TO VIBRATOR
                    BLACK                                   SOCKET PIN 2

                              FOR 6-VOLT VW'S       N.C.      PIN 4
                              FOR 6- VOLKSWAGENS, REMOVE
                              ONLY 60 TURNS OF EACH WIRE.
                    GREEN     USE HEP-232 TRANSISTORS
              BLUE            FOR Q1 AND Q2.
                                                            TO VIBRATOR
                                      GE-3                  SOCKET PIN 3
                              R3               Q2
                                                        2    1
MODIFIED STANCOR TA-16 TRANSFORMER   R4                        TOP
UNWIND SECONDARY - W/O DAMAGING COIL -                  3    4 VIEW
PERMITTING YOU TO DISCONNECT LIGHT-COLORED
WIRE FROM BLACK LEAD.  REMOVE 90 TURNS OF               VIBRATOR
THE LIGHT WIRE AND 90 TURNS OF YELLOW WIRE.             SOCKET
NOW CUT OFF ALL BUT ABOUT 4½" OF EACH WIRE.
RECONNECT LIGHT-COLORED WIRE TO BLACK.  RUN
YELLOW TRANSFORMER LEAD TO R1.
```

PARTS LIST

Q1, 2 — GE-3. For 6-volt sys‑
tems, use HEP-232
R1, 3 —240

R2, 4 — 11
T1 — See instructions on dia‑
gram. Use Stancor TA-16

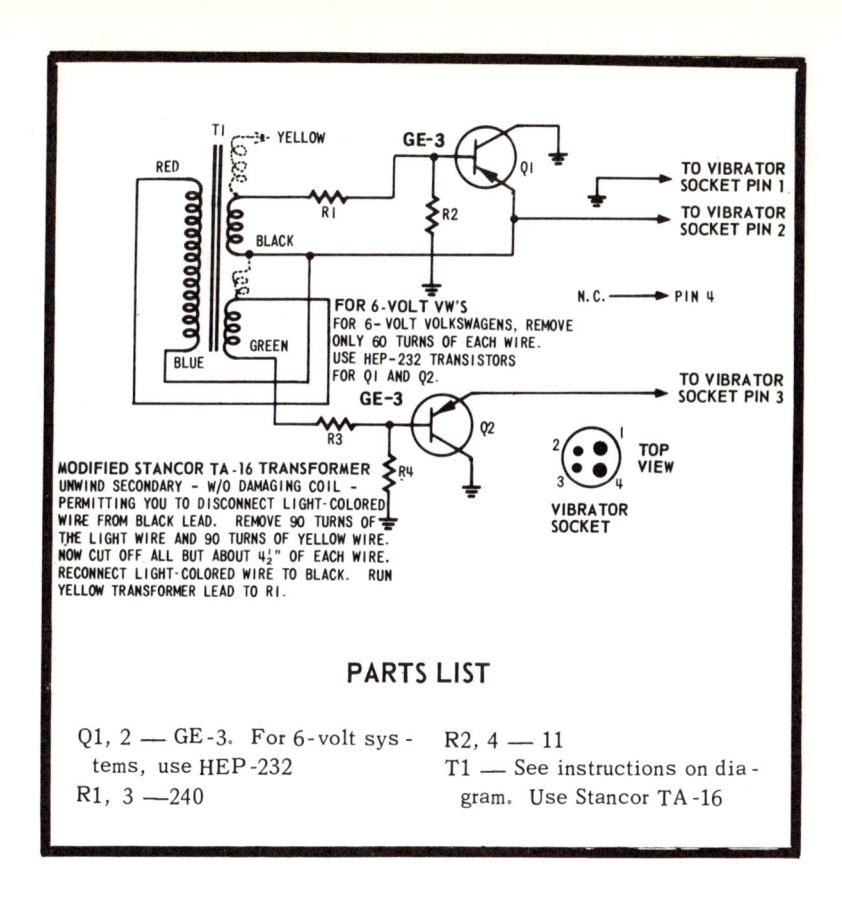

37

Junk Car Radio & Tape Player Retriever

Nothing is more frustrating than having to discard a perfectly good tape player or car radio because you don't have any way

OUTPUT

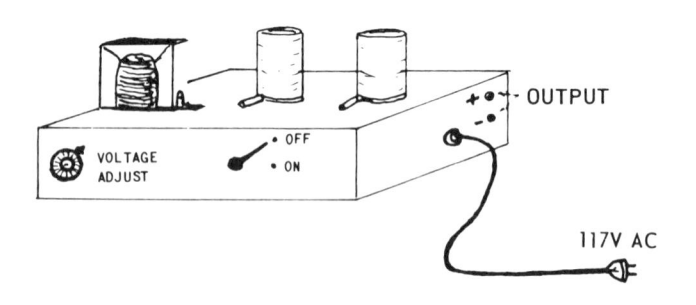

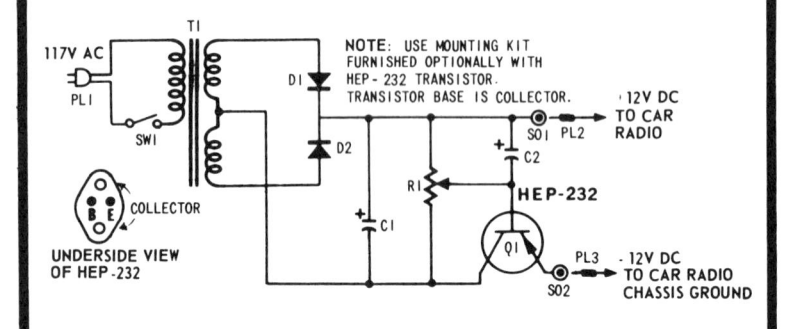

VOLTAGE ADJUST

OFF ON

117V AC

NOTE: USE MOUNTING KIT FURNISHED OPTIONALLY WITH HEP-232 TRANSISTOR. TRANSISTOR BASE IS COLLECTOR.

-12V DC TO CAR RADIO

-12V DC TO CAR RADIO CHASSIS GROUND

UNDERSIDE VIEW OF HEP-232

PARTS LIST

- C1, 2 — 1000 mfd, 50 WVDC electrolytic
- D1, 2 — 200 PIV, 12-amp silicon diodes. Lafayette SP-267
- PL1 — 117V AC power plug. Amphenol 61-F11
- PL2, 3 — Single-prong test plug. Amphenol 71-1S
- Q1 — HEP-232. Order with Motorola MK-15 mounting kit
- R1 — 2.5K pot
- SO1, 2 — Test jacks. Amphenol 78-1S
- SW1 — SPST, AC-household rating
- T1 — 117V AC primary; 12-29.8V AC secondary. Stancor RT-202

to make it work after the car that it was used in is gone. Your problems can be solved by this easy-to-build "retriever."

The project consists mainly of a transformer, two diodes, two electrolytic capacitors and a transistor. The transformer steps down 117 volts AC to a usable voltage for the circuit. The rectifiers (D1, D2) convert the AC into pulsating DC, which is smoothed out by the filter capacitors. A mounting kit, which can be purchased along with the transistor (HEP-232) at a small additional cost, should be used to mount it.

In this circuit, the values are not extremely critical, although they should be as close as possible to the specified ones. Be sure to connect the electrolytic capacitors as shown in the diagram, otherwise you will have real trouble on your hands. It never hurts to double check. You can adjust the output of your power supply with the variable resistor (R1).

38
VOM-To-Dwell Meter Converter

The dwell angle of your distributor is extremely important to the efficient operation of your car. Not only will the correct angle prolong the life of your points, but it will also keep the car's accelerating and hill climbing abilities up to par. An excessively large angle will burn up the points, while a small angle decreases the voltage induced in the coil, causing missing at high RPM. If you already have a voltmeter, you also have a potential dwell meter.

A modification circuit of four components will do the trick. Two resistors drop and adjust the voltage appearing at the meter. The zener diode keeps the voltage in the circuit rela-

133

PARTS LIST

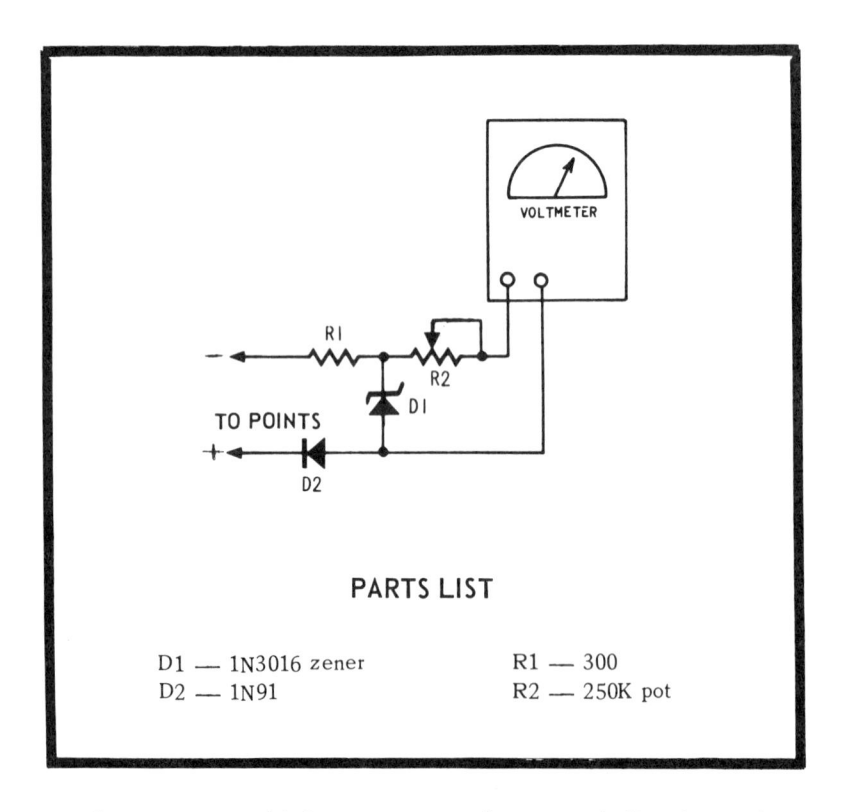

D1 — 1N3016 zener   R1 — 300
D2 — 1N91   R2 — 250K pot

tively constant, which means a steady meter indication. There is a directly proportional relationship between the average voltage and the dwell angle, making a linear scale possible. You may want to mount the elements in a small plastic case that can be attached to the meter for this particular function.

The test leads are red and black clips. To the side of the coil that is always connected to the distributor points goes the hot line. The other clip goes to a solid ground. The car ground is assumed negative. If it is positive, reverse the modification circuit connections to the meter and make the necessary adjustments on the meter to read negative voltage. Before using your dwell meter calibrate it on a known source. You may want to mark the different range settings on R2.

To use the meter, hook it up as already described. Start the engine and allow it to idle normally. You can read the dwell from the meter. For your automobile the manufacturer has specified the best dwell angle. In general, for eight cylinders it is $28°$, for six cylinders it is $35°$, and for four cylinders it is about $59°$.

39

Regulator Interference Killer

This simple trick will free you from interference created by the voltage regulator in your car. This interference, which you have undoubtedly noticed when listening to your car radio, is caused by the opening and closing of the relays contained in the regulator.

Connect this simple device—which merely consists of one resistor—across the generator field coil that is connected to the regulator.

This resistor acts as a filter, eliminating almost all of the troublesome static. For only the small cost of this resistor, you can be on the path to more pleasurable radio listening.

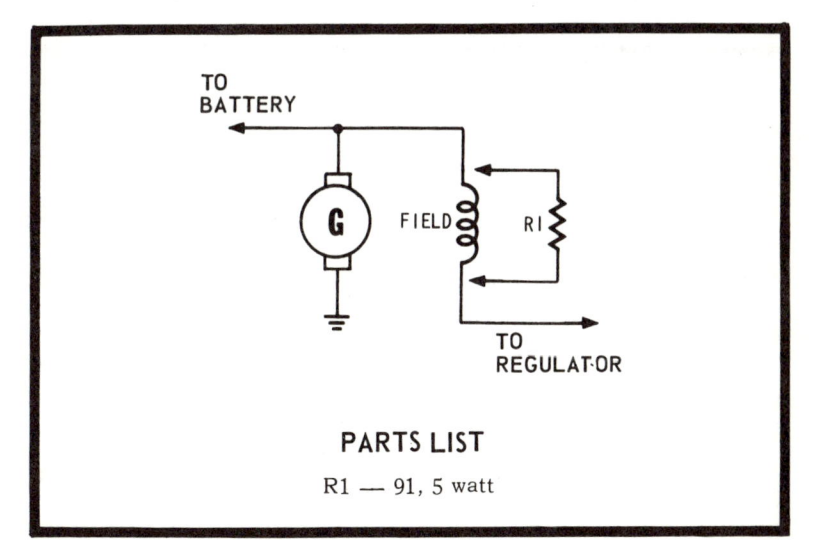

PARTS LIST

R1 — 91, 5 watt

40
The "Surpriser" Theft Device

Here is a marvelous gadget for protecting your automobile from theft. It consists only of an SPST switch that is connected from the coil side of the ignition to the horn circuit.

When you leave your car, turn the "surpriser" alarm on, making it impossible to turn the ignition on without connecting the horn circuit, too. If a prospective thief tries to start your car, he will probably be scared out of his wits by the sudden and continuous blast of your horn. If he doesn't immediately run and manages to find the alarm switch, he will have created a great enough disturbance to put himself in con-

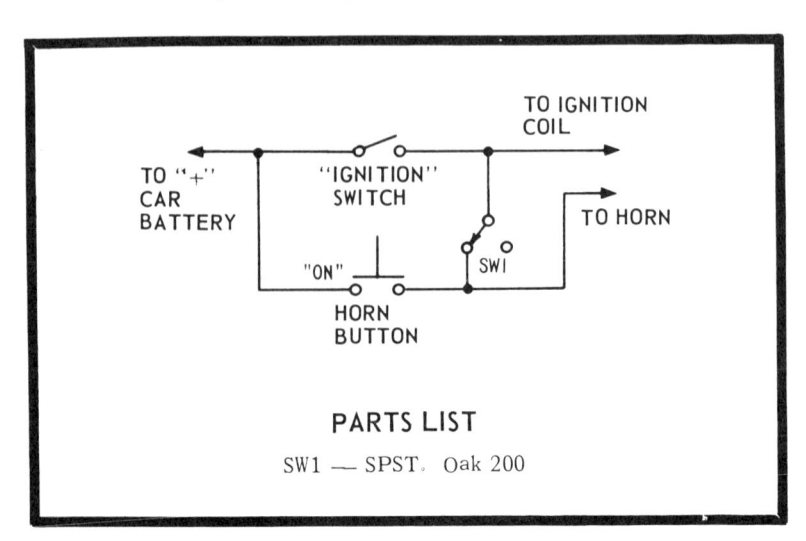

PARTS LIST

SW1 — SPST, Oak 200

siderable danger of being discovered—thus, safeguarding your car.

When you return to your car, merely turn the alarm switch off before turning on the ignition. Don't forget, though!

41
Transistorized Mobile Voice Control

Unless you have a switch on your mike, manual switching in mobile radio communications can be a real nuisance. It distracts attention that should be on the road. This project eliminates your having to think at all about switching, since the sole control mechanism is your voice. It is a 2-stage, voice-controlled relay with the relay connected into the transceiver control circuit. Merely speak into the mike and the transmitter is on; cease talking and you're back on "receive," ready for the other fellow to begin.

The core of the project are two transistors, transformer-

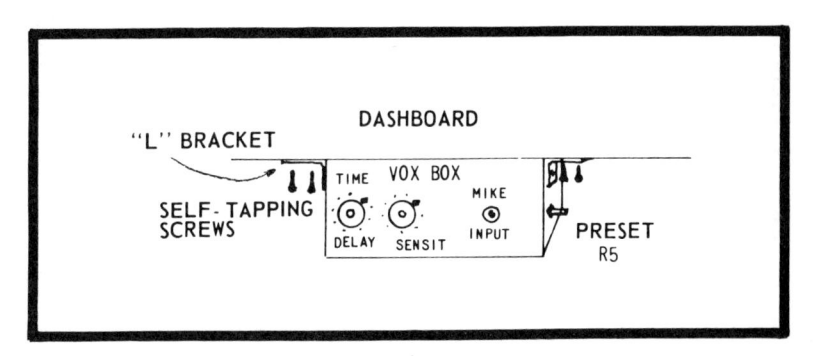

PARTS LIST

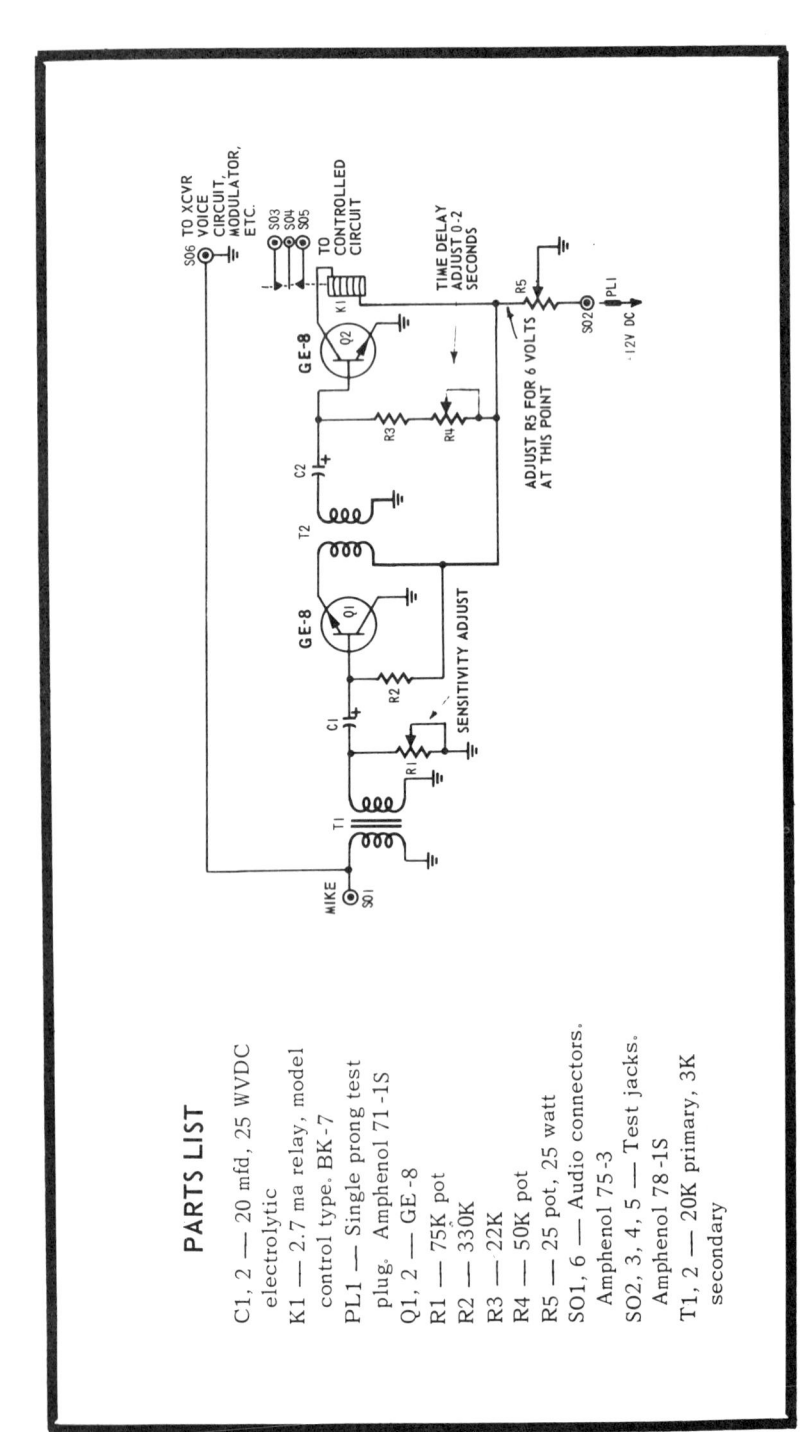

C1, 2 — 20 mfd, 25 WVDC electrolytic
K1 — 2.7 ma relay, model control type. BK-7
PL1 — Single prong test plug. Amphenol 71-1S
Q1, 2 — GE-8
R1 — 75K pot
R2 — 330K
R3 — 22K
R4 — 50K pot
R5 — 25 pot, 25 watt
SO1, 6 — Audio connectors. Amphenol 75-3
SO2, 3, 4, 5 — Test jacks. Amphenol 78-1S
T1, 2 — 20K primary, 3K secondary

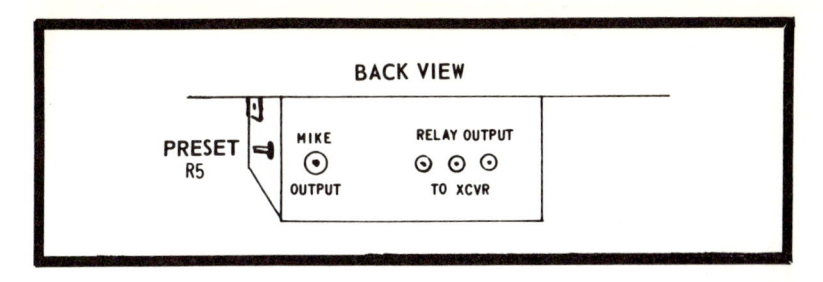

```
┌─────────────────────────────────────────────────┐
│                   BACK VIEW                      │
│                                                  │
│              MIKE      RELAY OUTPUT              │
│   PRESET      ⊙         ⊙ ⊙ ⊙                    │
│     R5      OUTPUT       TO XCVR                 │
└─────────────────────────────────────────────────┘
```

coupled in two amplification stages, where the minute audio voltage from the mike is boosted to the point where it will trip the relay. Preferably, the break into the mike line should be made very near or inside of the transceiver. Shielding for your circuit is not critical since it is not at all responsible for carrying the audio intelligence. The only danger is the coaxially shielded mike cord itself, which should be isolated from the bulk of your components. A small metal box, attached under the dashboard along with the radio, is best for construction. It will, however, have to be large enough to house two coupling transformers and a relay.

How fast the circuit responds to your voice is extremely important. This is determined by R4, the time-delay adjust. If you set the delay too short, the relay will be cutting out between your every word. You will want to set it so that you can pause after a word and the relay will not kick out before you start the next one. This means a slight delay when you initially begin speaking, but this should be no problem. Clear your throat before you start talking, or something. You will have to adjust R4, keeping in mind your own talking habits and mannerisms.

R1 determines the sensitivity of your voice-controlled circuit. Set it to trip the relay at your normal speech level.

Your car battery is the source of power of your circuit. The circuit does, however, need only six volts, the remaining six being dropped across potentiometer R5, which must be adjusted for the correct voltage drop. The relay, replacing the manual switch, is connected right into the control circuit of your set. No longer will you be plagued by the potentially dangerous distraction of an on-off manual control.

42
Inexpensive 117-Volt AC Inverter

Today, there are many luxury appliances available such as televisions, radios, and electric razors that operate on both AC and self-contained, rechargeable batteries. These units are perfect for traveling. However, what about pieces of

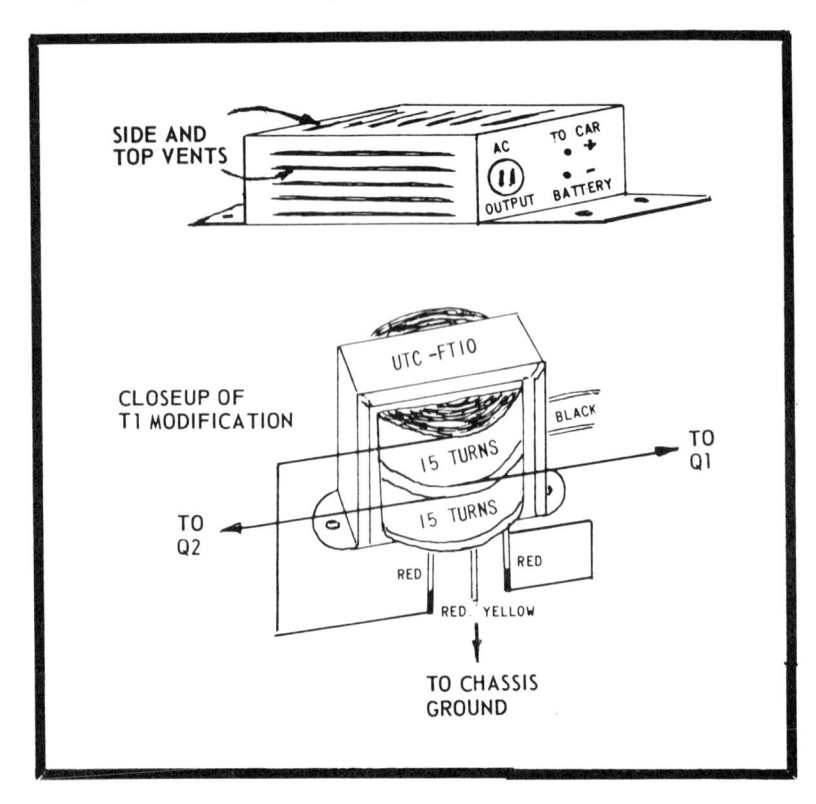

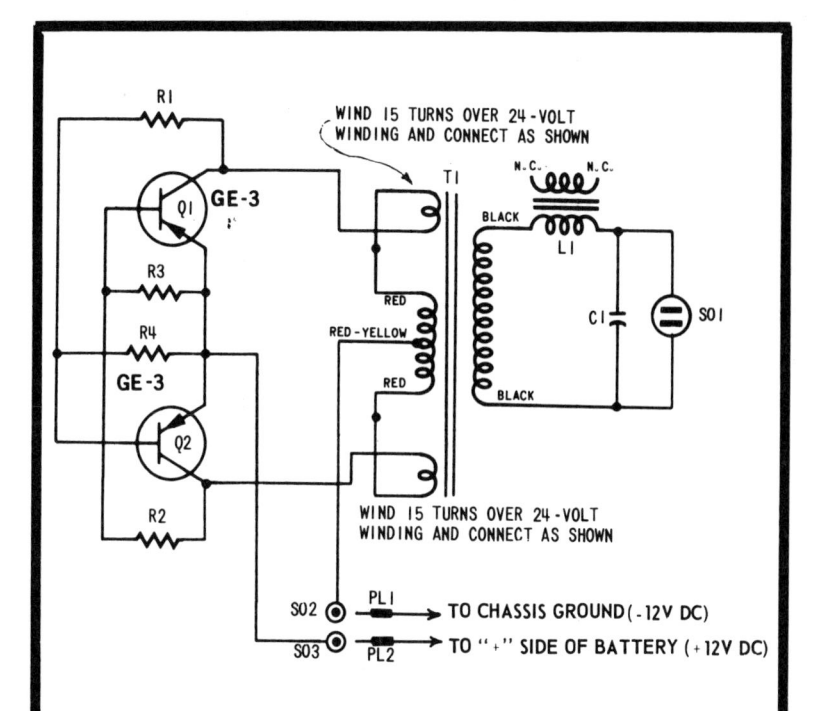

PARTS LIST

C1 — 2 mfd
L1 — 6.3V AC fil transformer, Stancor P-6134. 117V AC primary is not used
PL1, 2 — Single-prong test plugs. Amphenol 71-1S
Q1, 2 — GE-3
R1, 2 — 110
R3, 4 — 16
SO1 — AC socket. Amphenol 61-MIP-61F
SO2, 3 — Test jacks. Amphenol 78-1S
T1 — 24V AC fil transformer with CT secondary, UTC FT-10. Make modifications shown in diagram, winding two 15-turn coils over outside of transformer.

equipment, like slightly older units, that do not have this convenience? They can still be used in the car or boat by building this simple and inexpensive inverter. This unit not only increases the voltage but also provides the AC action that is needed to run normal AC equipment.

The two transistors, Q1 and Q2, form the basis of operation of this device. They are hooked up in the circuit so that they perform a very rapid and sustained switching action, which in turn produces the desired alternating current. However, the constant switching action causes the transistors to become quite hot, necessitating the use of heat sinks.

The other main component—the transformer—has to be modified slightly before being installed. The transformer has to boost the 12-volt output of the battery to the required 117 volts. This means using different secondary windings. These can be constructed by winding two identical coils on the outside of the transformer. Wind and connect them as shown in the diagram. Notice that the 117v AC primary coil is not used. Make sure that all connections are of a heavy-duty nature and are properly insulated to avoid dangerous shocks.

The unit should be constructed in a heavy metal container which should be well insulated. This includes using rubber grommets and heavily insulated wire and cable. Heat sink the transistors, ventilate the container and mount a normal AC wall receptacle on the front of the unit. When the unit is completed mount it in the car or boat permanently, tapping into the positive line running from the battery. In a car, the input wires can be run into the passenger compartment along with the other wires running to the dashboard. The unit should be solidly installed on the drive shaft hump under the dashboard. This can be done by shaping two brackets out of iron strapping so that they fit the slope of the hump. Firmly attach the brackets to the hump with self-tapping screws. These brackets then provide a solid base on which to mount the completed unit.

In use simply plug in the television, radio, or other appliance and you are ready to go. This inverter will power dozens of different objects; however, try to avoid using it for heavy-current appliances such as larger tube-type television sets, etc. Be sure to use it only while the engine is running so that you do not wear down your battery. As long as the engine is running, the generator or alternator will keep the battery charged. Once the engine is turned off, though, it does not take long to exhaust the battery.

43

Auto Ice Alarm

Here's a dandy gadget which will let you know when the outside temperature approaches the freezing mark. So what? Simply because that when the 32° mark is hit, driving often becomes hazardous due to icy spots in the roadway. In essence, then, this is a "CAUTION" sign that can help you prevent having an accident.

The project consists of three transistors and a "light-alarm" bulb which flashes when freezing temperatures occur outside the automobile. Heart of the gadget is the 2N397 transistor

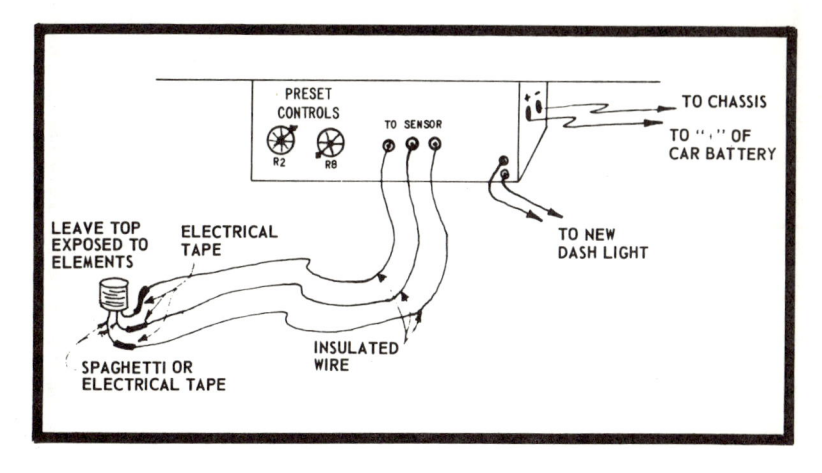

143

PARTS LIST

C1 — 25 mfd, 25 WVDC electrolytic
I1 — 6-volt pilot bulb
Q1, 2 — 2N397
Q3 — GE-2
R1 — 11
R2 — 500 pot, 2 watt
R3 — 230
R4 — 6.9K
R5 — 16K
R6 — 1.1K
R7 — 210, 2 watt
R9 — 160, 2 watt
SW1 — SPST, Oak 200
SO1, 2, 3, 4, 5, 6, 7 — Test jacks, Amphenol 78-1S
PL1, 2, 3, 4, 5, 6, 7 — Single-prong test plugs, Amphenol 71-1S

mounted in the grillwork up front. This senses temperature change, alerts the main circuit, and causes the bulb to light.

You could mount the main circuit in the glove compartment, with only the 2N397 (Q1) in the grill. It can be taped in position or held in place with a small piece of a youngster's modeling clay.

Be sure to "calibrate" the instrument according to instructions on the diagram. Settings are not critical, but will be a guide for the circuit to follow. Be sure to adjust R8 so that it kicks out somewhere near 36°. Otherwise, it'll keep flashing forever.

44
250-Volt Mobile Power Supply

Ever since Detroit auto makers started installing transistorized radios in cars, it has been a problem finding the 250-volt output needed to power the tubes in some mobile ham or CB units. However, this doesn't have to be the case. With this mobile power supply you can use your equipment in even the newest cars. This unit takes the 12-volt output of the battery and boosts it to the required 250 volts needed to power the tubes.

In this seemingly simple operation there are actually three separate phases involved. First, to get the transformer to operate, the DC must be "chopped" into an alternating current. This is primarily what the transistors do. With their collectors grounded the circuit becomes an oscillating circuit, turning the direct current into AC current of equal voltage. Next, this oscillating current is fed into the transformer coil

PARTS LIST

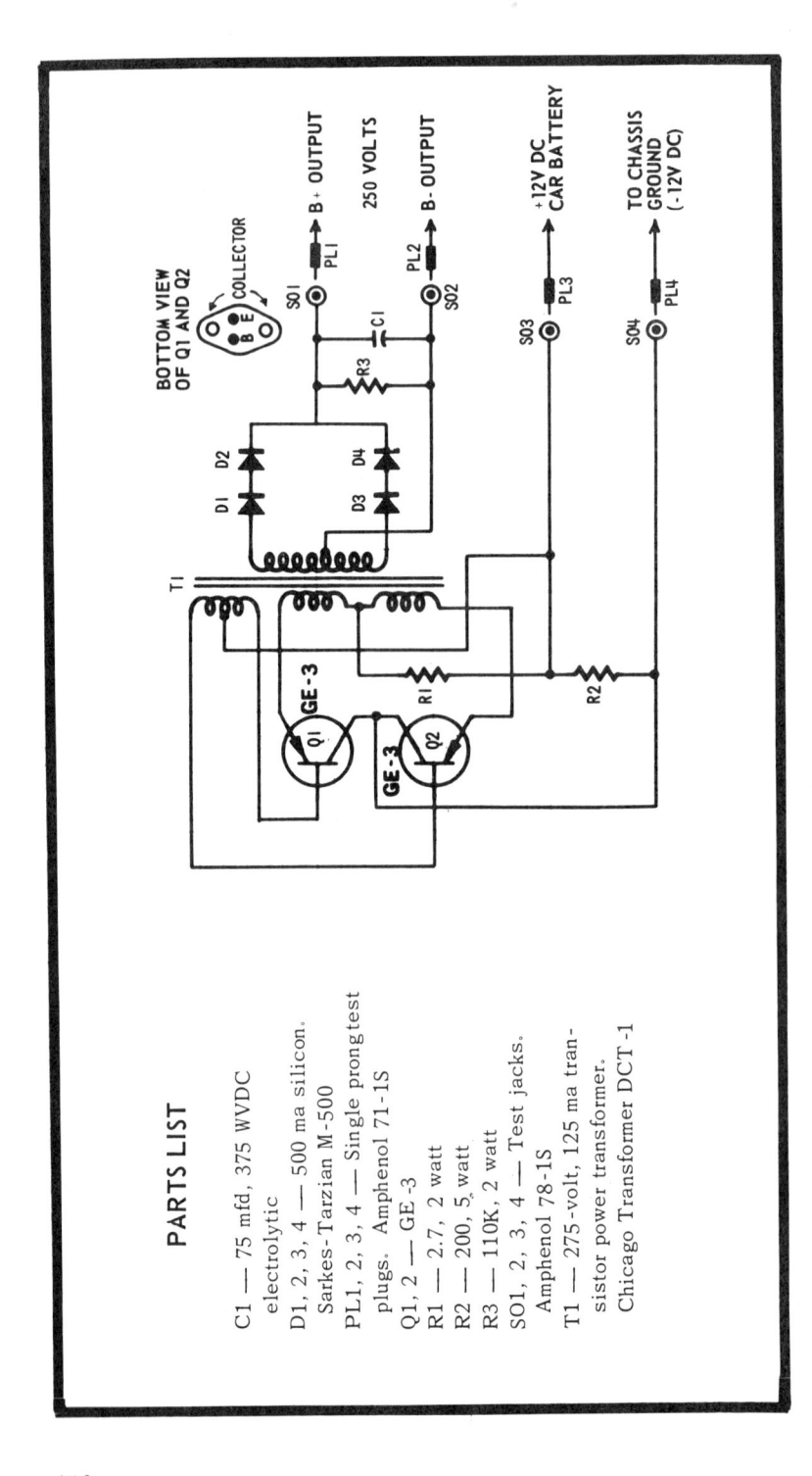

C1 — 75 mfd, 375 WVDC electrolytic
D1, 2, 3, 4 — 500 ma silicon. Sarkes-Tarzian M-500
PL1, 2, 3, 4 — Single prong test plugs. Amphenol 71-1S
Q1, 2 — GE-3
R1 — 2.7, 2 watt
R2 — 200, 5 watt
R3 — 110K, 2 watt
SO1, 2, 3, 4 — Test jacks. Amphenol 78-1S
T1 — 275-volt, 125 ma transistor power transformer. Chicago Transformer DCT-1

where the voltage is stepped up. The next step again changes the form of the current. The output from the secondary winding is then fed into a battery of junction rectifiers. The current loses its oscillatory motion and again becomes positive and negative direct current, although now with a potential of 250 volts.

In constructing this unit, although it is very simple, care must be taken in making sure that a few small details are watched closely. For instance, the transistors have to be connected in the circuit as shown, with the emitters connected to one of the primary windings and the two bases connected to the other primary winding. Just be careful to match the bottom of the transistors with the drawing and make the connections right the first time since they can be irreparably damaged if they are connected wrong and the power is applied. Also, the rectifiers have to be installed with the proper bias so that the current will flow in the output.

Construct the device in a heavy metal container, making sure there is ample ventilation for the two transistors which must also be used with heat sinks (see pictorial, Project 52). Use all heavy-duty materials, including heavily insulated wire and cable. Rubber grommets and electricians' insulating tape will enable you to thoroughly insulate the unit to avoid any nasty shocks.

Once the device is constructed it can be installed in either car or boat. In a car run the tap line from the positive lead of the battery and the ground wire through the firewall where the other dashboard connecting wires pass. The power supply can be suspended from under the dashboard using perforated hanger iron. Finally, just connect the output of the power supply to whatever mobile unit you are using.

45
Subminiature Tachometer

Proper setting of the carburetor is extremely important to the operating efficiency of your automobile. With this simple tachometer you can accurately determine the optimum gas-air mixture and the correct idling speed for your car. You will find your car running much better, not only in terms of gas mileage, but also in acceleration potential. Also, by belching out less unburned gas into the atmosphere, you will be doing your part in curbing air pollution!

The circuit itself is relatively simple, consisting primarily of two transistors and a milliammeter. A small metal box that will conveniently house a meter and the other components in the circuit is used, and construction can be similar to that in Project 1. Since there is no need to isolate the components to avoid short circuits from switching grounds, the chassis is ground.

During operation, the tachometer picks up voltage pulses generated by the distributor points opening and closing according to engine speed. Only the AC component of the pulses gets through capacitors C1 and C2, which block the DC potential that the transistors cannot handle. As the RPMs increase, the pulses become narrower and tend to vary in amplitude.

It is consequently necessary to "clip" the pulses, only the frequency being important. The 2-transistor combination does this and also provides the amplification necessary to register the signal on the meter. A zener diode also minimizes voltage fluctuation. The meter indicates the average value of the incoming pulses. If the number of pulses is doubled, so is the average voltage reading. This makes it possible to read the RPM directly from the linear scale.

To use the tachometer, hook your test lead to the distributor points terminal and connect the battery. Ground the meter chassis and negative battery terminal well. Although shock is no problem with the battery connections, the spark plug cables are dangerous to the person tinkering with the engine while it is on. Your violent recoil from these relatively mild voltages can throw your hand unwittingly against the fan or a hot pipe. In other words, make the meter connections while the engine is off.

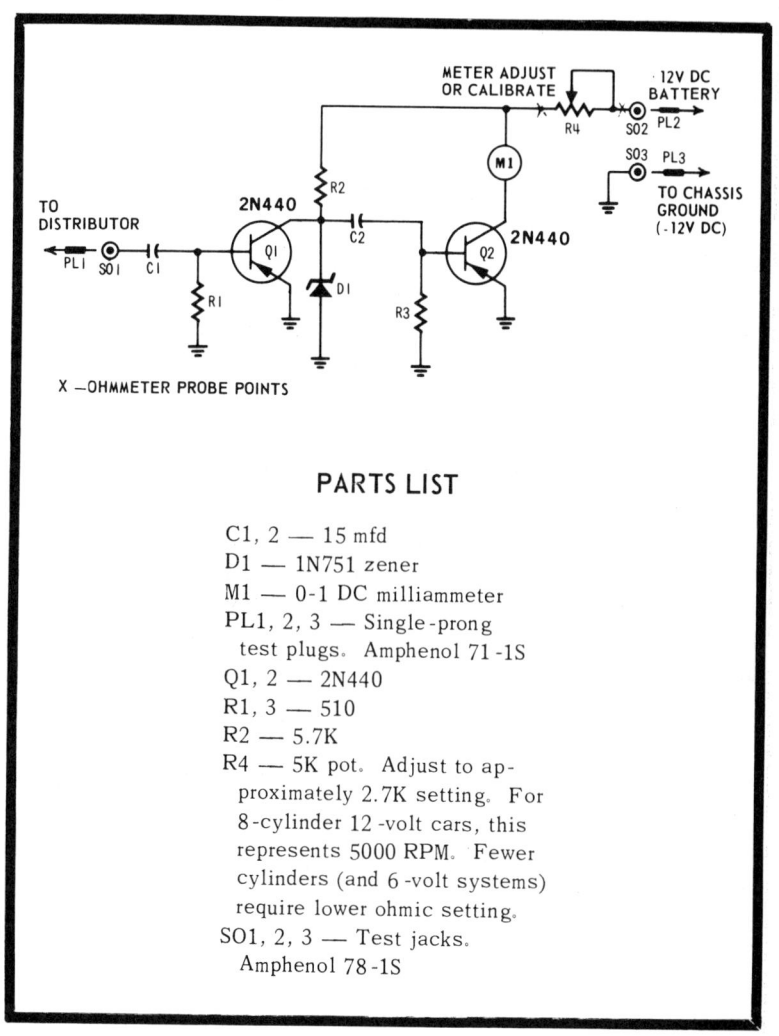

X —OHMMETER PROBE POINTS

PARTS LIST

C1, 2 — 15 mfd
D1 — 1N751 zener
M1 — 0-1 DC milliammeter
PL1, 2, 3 — Single-prong test plugs. Amphenol 71-1S
Q1, 2 — 2N440
R1, 3 — 510
R2 — 5.7K
R4 — 5K pot. Adjust to approximately 2.7K setting. For 8-cylinder 12-volt cars, this represents 5000 RPM. Fewer cylinders (and 6-volt systems) require lower ohmic setting.
SO1, 2, 3 — Test jacks. Amphenol 78-1S

Your tachometer is designed to peak in the range of 5000 RPM, plus or minus 2000 RPM, although it is reliable for a larger range. Resistor R4, a potentiometer, in conjunction with a VTVM, enables you to calibrate the meter, with an ohmic reading of about 2.7K for a 12VDC, 8-cylinder car. The circuit will operate also in 6-volt and in systems with fewer cylinders, but a lower setting of R4 is required. Calibrate your tachometer against a known source. Experiment to obtain your meter's working conditions for your car.

Although the tachometer can be used for several automotive engine tests, its most common use is the setting of carburetors. To adjust a carburetor, set the idle mixture screw for the greatest RPM reading. If it has more than one mixture screw, each one should be adjusted until the engine reaches its highest RPM. Then adjust the idle speed until the engine idles at the correct speed. You will want to repeat the procedure several times to insure perfection.

You can further check your air-fuel ratio by reducing idle speed to 800 and then 1500 RPM. Remove the air cleaner and slide a flat plate slowly over the intake so as to partially choke off the flow of air. The tachometer should be watched closely for change. An increase in speed of as much as 50 RPM means the mixture is normal; no increase, the mixture is too lean.

Engine timing can be optimized with the tachometer, too. Loosen the distributor cap and rotate the distributor until the highest steady RPM is obtained. Retard the spark (rotate in the direction of rotation) until the speed drops about 40 RPM and tighten the cap. (Timing should be checked with a strobe light.) The farther the spark is advanced without causing pinging, the more efficiently the engine will run.

Spark plugs can be tested by letting the engine idle in a normal manner and shorting out each plug one at a time. If the engine speed decreases, the plug is good; if the speed remains the same, the plug is not firing.

A wavering of the meter pointer at a high RPM means that the distributor points are not operating efficiently. A comparison of speedometer and tachometer while the car is accelerating will reveal the condition of your clutch. With the rising cost of car maintenance, you cannot afford not to own a tachometer—the universal tune-up instrument.

46
Simple Dwell Meter

The dwell or cam angle of a car is the number of degrees that the points remain together during the entire ignition cycle. The points are spaced too close if the dwell angle is high and are too wide if the angle is small. A high dwell allows excessive current to flow through the coil and points, causing the point contacts to burn. A small dwell may be responsible for poor performance at high speed, causing the engine to "miss" while accelerating or hill climbing, because the angle does not permit the coil to build up a sufficient magnetic field and, therefore, reduces coil voltage.

If you have had to replace your points frequently, or if your car does not accelerate as well as it should, you will find this simple, easy-to-build dwell meter a worthwhile investment, both in the new-found efficiency of your car, and in the long-run savings of do-it-yourself maintenance work.

Build the circuit in a Minibox, or a container designed to accommodate a small meter in one of the sides. (See pictorial, Project 1.) The meter is operated on two batteries, making it completely portable. Watch the polarities of C3, an electrolytic, and the rectifier. Switches SW1 and 2 offer a certain amount of variation in the control of the dwell angle being measured, along with potentiometer R1, which serves as a gain control. The switches merely add another "smoothing" capacitor to the circuit to minimize fluctuation in the pulsating DC passed by the rectifier.

The output leads should be distinguishable by color—one red, one black. The hot one goes to the coil terminal which is always connected to the distributor points. The other lead

151

PARTS LIST

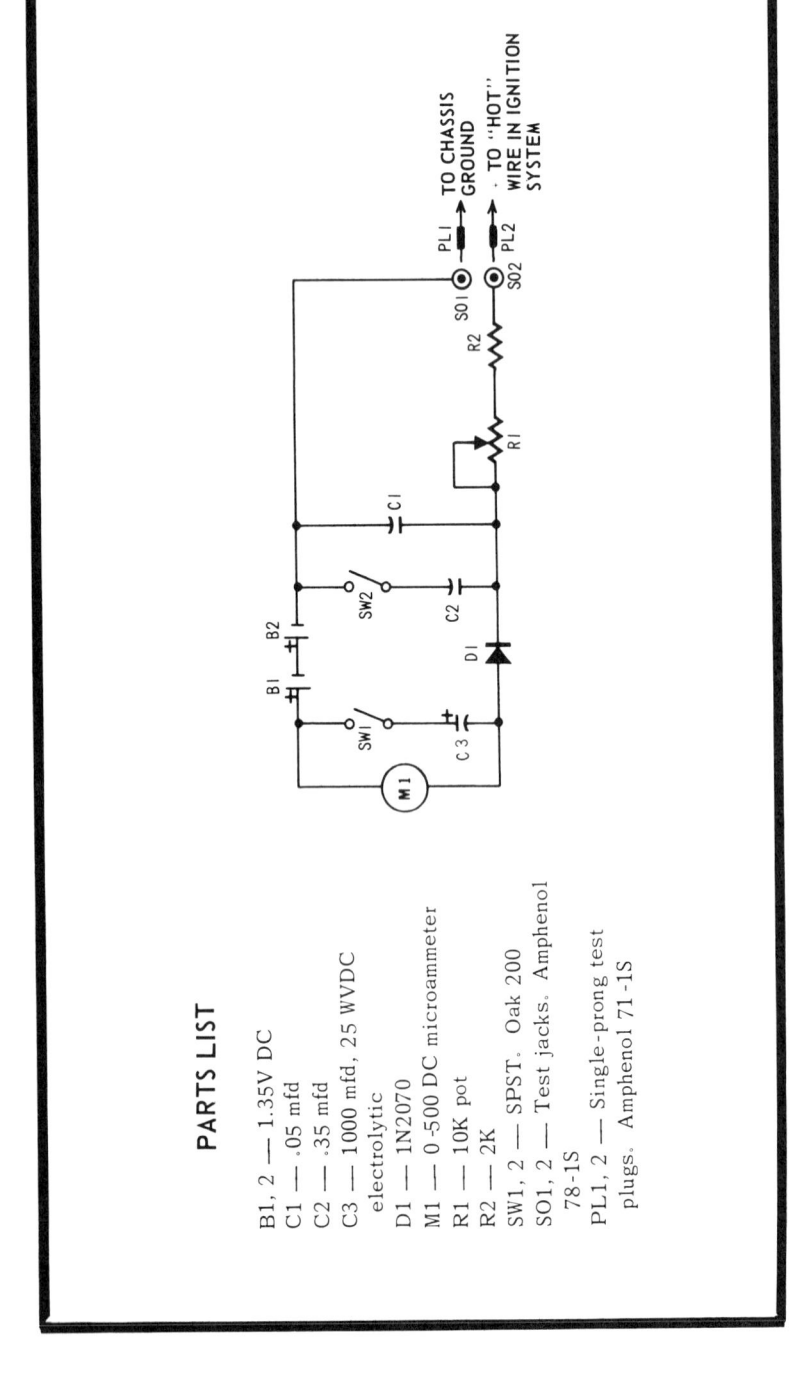

B1, 2 — 1.35V DC
C1 — .05 mfd
C2 — .35 mfd
C3 — 1000 mfd, 25 WVDC electrolytic
D1 — 1N2070
M1 — 0-500 DC microammeter
R1 — 10K pot
R2 — 2K
SW1, 2 — SPST. Oak 200
SO1, 2 — Test jacks. Amphenol 78-1S
PL1, 2 — Single-prong test plugs. Amphenol 71-1S

goes to a good ground. The meter will have to be carefully calibrated against a known source initially.

To find the dwell angle, start the engine and permit it to idle in a normal manner. Read the dwell angle from your calibrated meter. To adjust the points, connect the meter as described above. Turn on the ignition and crank the engine by means of the starter motor. While observing the meter, adjust the points to the manufacturer's recommendations. This angle is about 28 for 8 cylinder cars, 35 for 6 cylinder, and 59 for 4 cylinder. Recheck the angle with the engine running; in other words, when it has reached normal operating temperature. The reading is apt to be slightly higher when the engine is running due to the starter motor current.

89c Direction-Finding Antenna 47

Here is an excellent home-brew highly directional antenna that will pinpoint the direction of even a relatively distant signal source. It also will serve you well for fox hunts, tracing down local sources, or just about anything that you might have in mind. All you need is a section of window screening, a length of flexible copper tubing, and a piece of scrap lumber.

The heart of the project is a loop of copper tubing approximately 1/8 wavelength long, which is about 9-1/2" at 144 MHz. The loop is connected in the same way as a dipole to the length of coaxial cable running down to the receiver—one lead to the shielding, the other to the center conductor. Around the loop is a cone of window screening about 1/4 wave-

153

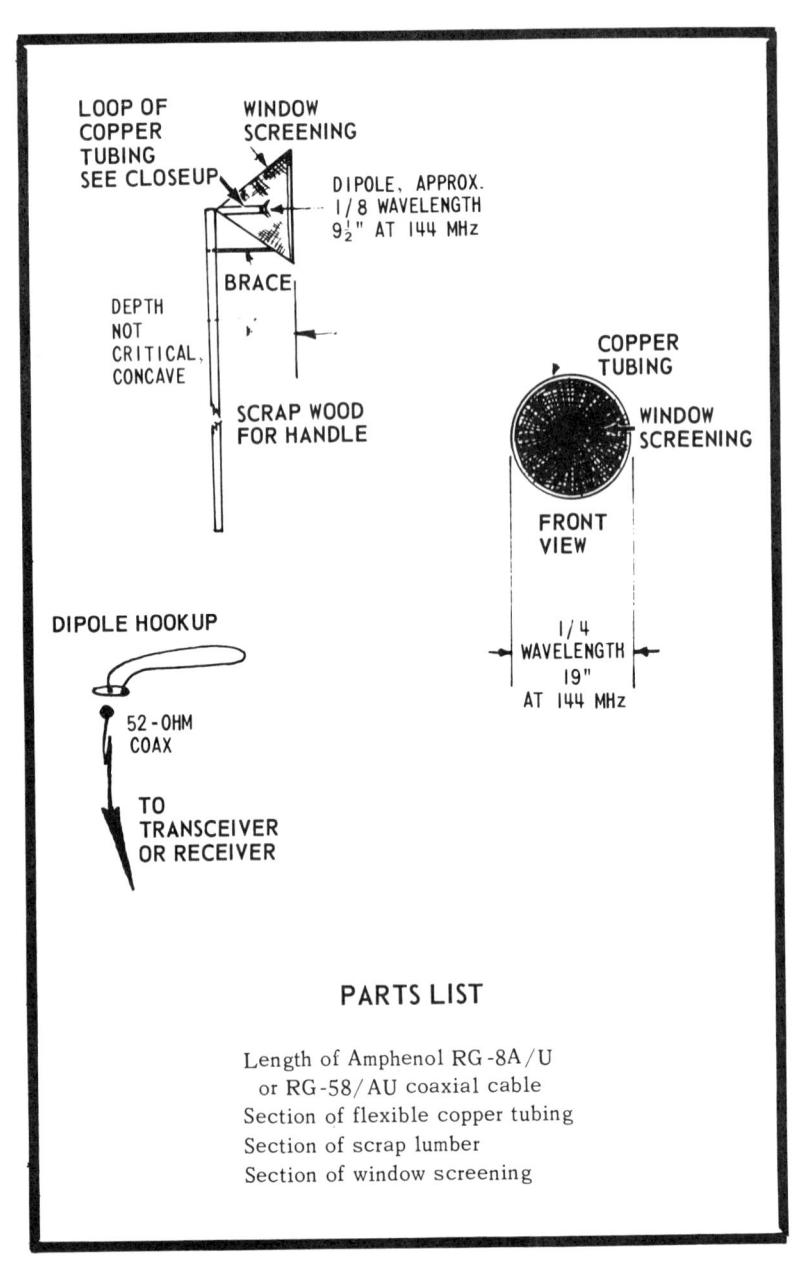

```
LOOP OF        WINDOW
COPPER         SCREENING
TUBING
SEE CLOSEUP
                    DIPOLE, APPROX.
                    1/8 WAVELENGTH
                    9½" AT 144 MHz

        BRACE

DEPTH
NOT                                    COPPER
CRITICAL,                              TUBING
CONCAVE
                                            WINDOW
                                            SCREENING
        SCRAP WOOD
        FOR HANDLE

                                       FRONT
                                       VIEW

DIPOLE HOOKUP
                                         1/4
                                      WAVELENGTH
                                         19"
        52-OHM                       AT 144 MHz
        COAX

        TO
        TRANSCEIVER
        OR RECEIVER


                    PARTS LIST

            Length of Amphenol RG-8A/U
              or RG-58/AU coaxial cable
            Section of flexible copper tubing
            Section of scrap lumber
            Section of window screening
```

length in diameter, which is about 19" at 144 MHz. A stabilizing ring of copper tubing gives form to the screen cone (see pictorial). The depth of the cone matters little, as long as it provides adequate shielding from the side and back.

The cone and dipole—which do not touch— are mounted on a wooden pole and are supported by a short wooden brace. The coax runs conveniently down the pole.

Mount the antenna on the side of your car with plenty of elbow room for free turning. The antenna becomes ungainly for lower frequencies, being primarily useful for upper VHF.

48

One Transistor Auto Light Minder

Driving in the daytime with lights on is generally not considered to be much of a problem. It can, however, cause considerable trouble. Since you're not aware that your lights are on, you may leave the car without turning them off, in all likelihood, exhausting your battery to the point where your car will be "dead" too. Short trips with the lights on are always bad, day or night, because of insufficient recharging time. This auto light minder will alert you to the fact that you have your lights on. You will at least have the choice of whether to turn them off or not.

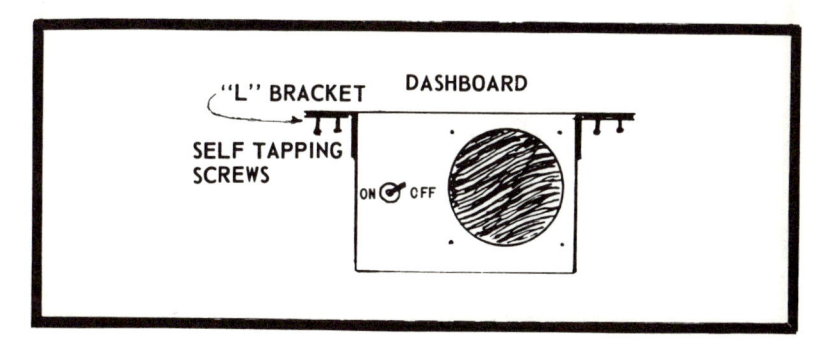

BACK VIEW

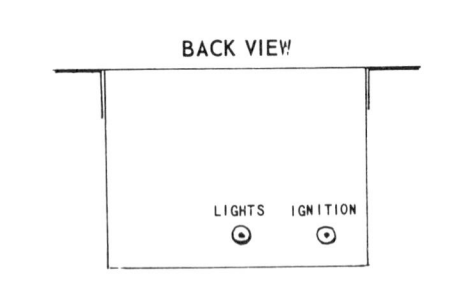

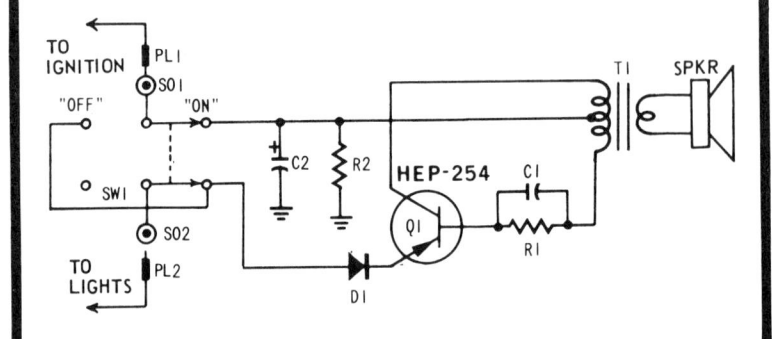

LIGHTS IGNITION

PARTS LIST

C1 — .25 mfd
C2 — 25 mfd, 25 WVDC electrolytic
D1 — 1N38B
PL1, 2 — Single-prong test plugs. Amphenol 71-1S
Q1 — HEP-254
R1 — 16K
R2 — 750
SO1, 2 — Test jacks. Amphenol 78-1S
SW1 — DPDT. Oak 200
T1 — 400-ohm center tapped primary; 11-ohm secondary (audio input transformer)

The minder should be constructed in a small metal box and mounted out of sight under the dashboard, with only your on-off switch exposed. One side of the DPDT runs to the ignition, the other to the light circuit. The audio output transformer goes directly to the speaker of the car radio. When power derived from the completed ignition-light circuit is applied to it, the light minder becomes an oscillator, producing a loud, clear tone in the speaker—until the lights are turned off, that is. For night driving, merely switch your alarm off.

49
Powerful In-Car PA System

Here is a handy circuit that will really sock it to anyone in the near vicinity of your car. It is a forceful 2-transistor PA system that can be useful at picnics, parades, or even in civil defense activities. Whereas most systems of this type make use of the car radio's audio amplifier, this one is completely independent of the car radio.

The microphone can be an old carbon earphone, since a voltage is applied to it in the circuit. The whole thing is mounted in a small box and stuck under the dashboard, installed with a small hook for the mike. R2 is the gain control and is combined with the on-off switch for convenience. 12 volts from the car battery completely satisfies power requirements. The negative ground is the car chassis.

From the little box under the dash run two wires to the front grill of the car where the speaker is mounted. This speaker is preferably the horn type, which makes the most effective use of the available power and is weatherproof to

PARTS LIST

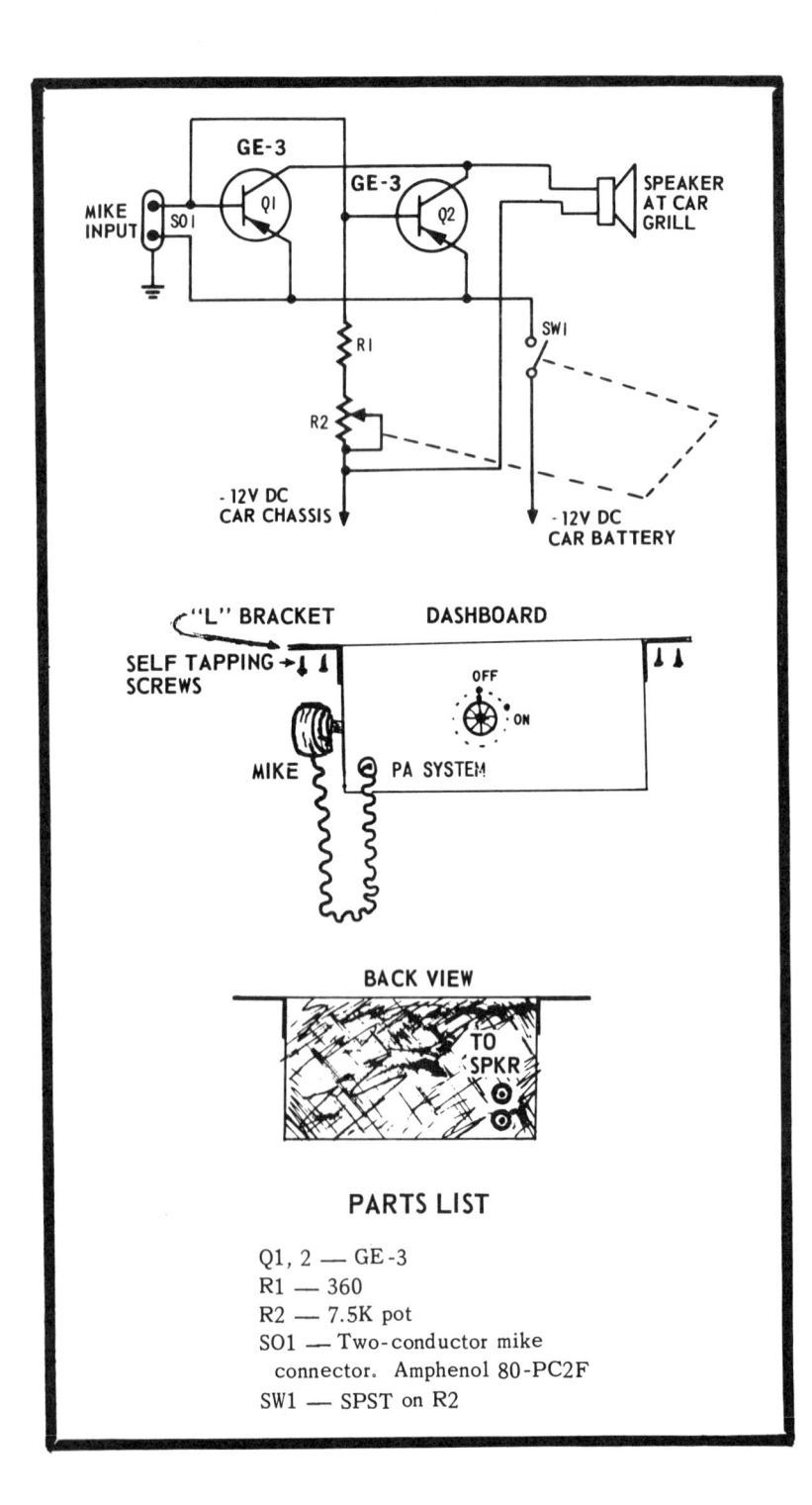

Q1, 2 — GE-3
R1 — 360
R2 — 7.5K pot
SO1 — Two-conductor mike connector. Amphenol 80-PC2F
SW1 — SPST on R2

boot. The paper diaphram type will work also, covering a reasonably large area. It is entirely possible to consider entertainment aspects of your PA, amplifying the minute audio detected by your car radio and shooting it through the speaker on the grill.

50
Inexpensive Battery Charger

A handy battery charger is probably the most indispensible piece of automobile maintenance equipment you can have. Careless "slips" or just routine driving can create a dead or ailing battery. Short runs at night with headlights and other electrical equipment on may result in a run-down battery because of insufficient recharging time to restore it to operating capacity. Repeated attempts to start an ill-tuned car can also lead to trouble. Parking lights left on all night usually don't do the battery any good either! In addition, the

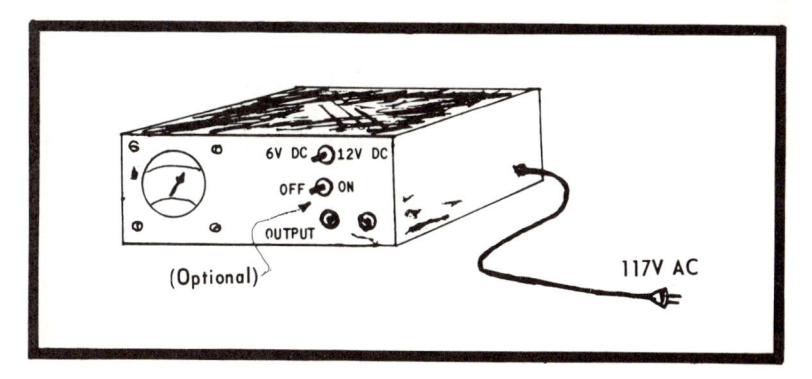

PARTS LIST

C1 — 100 mfd, 25 WVDC electrolytic
D1, 2, 3, 4 — 50 PIV, 2 amp silicon. Allied 39A720-D
F1 — .1A 3AG, Littelfuse
I1 — NE-51 neon
M1 — 0-100 DC milliammeter
PL1 — AC wall plug. Amphenol 61-F11
PL2, 3 — Single-prong test plug. Amphenol 71-1S
R1 — 110K
SO1, 2 — Test jacks. Amphenol 78-1S
SW1 — DPDT. Oak 200
T1, 2 — 6.3V AC filament transformer. Triad F-14X

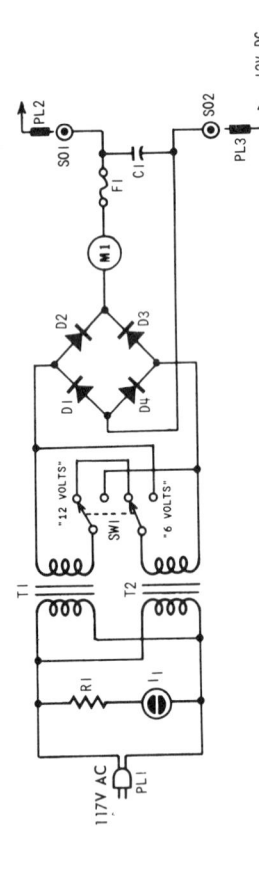

output potential of a battery is drastically reduced as the temperature drops further below freezing. An overnight charge will often prevent the below-zero morning catastrophy of an inanimate auto.

This charger is little more than a step-down power supply designed for relatively high current (100 ma max). 117 AC is brought down to 12 or 6 volts depending upon the combination of the two secondary coils. Switch SW1 is the selector, your choice depending, obviously, on whether you're charging a 6 or a 12-volt battery. A full-wave diode bridge rectifier and single large capacitor provide a relatively smooth DC output. A very low battery may take several hours to recharge, while a battery exhausted by repeated attempts to start a car may be rejuvenated in merely an hour. The current tolerance of the charger is not high, and it is best used for only mildly run-down cells.

A small metal box, with the current meter, selector switch, and terminals mounted on the front, is ideal for construction of the circuit. If desired, a single-pole single-throw switch in one leg of the primary circuit can be used to turn the charger on and off. A neon lightbulb will let you know when it's on. The normal precautions against an accidental ground are especially important here. Be sure to get the polarity of the electrolytic capacitors and the diode right. For the charger leads it is advisable to use the traditional red and black clip leads to keep them straight and avoid switched connections.

To use the charger, hook the leads to the battery first—positive to positive, negative to negative, then plug in the charger. Be very careful not to switch the connections, which would cause a direct short to ground and may ruin your charger even before you get started. If the car contains an alternator, this will also be damaged to the extent of requiring replacement. The meter will indicate a current flow through the battery. Do not allow the meter arm to go off the scale. You will find the limit by testing it on batteries in various stages of being rundown. While charging, remove the cell caps to release excess gas. The gas (hydrogen) is the result of hydrolysis and is extremely flammable.

When the meter reading reaches zero and a hydrometer reading reaches the correct point, the charging is finished. It is best to charge the battery only to the point where the car will start. Much more charging current is supplied by the running engine itself than can possibly be supplied by your inexpensive battery charger.

When you anticipate trouble starting on a cold morning, hook the charger up the night before. It will provide the reserve that you need to get started.

51
Soldering Gun For Your Car

Ordinarily, if you're on the road, the only way to remedy a loose wire is to open your pocket book at the nearest gas station. A traveler who has found the cause of the problem, but is at a loss for a way to fix it without a soldering iron finds himself confronted with the same dilemma, a waste of time and money. But what about a portable iron? Why should you be any less equipped to handle an electrical problem in your car than you are to take care of a flat tire? The "home-brew" iron described here is operated from the car battery and can easily be tucked under the seat or into the trunk of your car.

The components for the iron can surely be rummaged from your downstairs workshop junkpile. If not, they are very inexpensive and can be picked up at any hardware store. The handle from an old file is a good place to start. Into the handle is inserted a length of 5/16" diameter copper tubing. The length of the tubing does not matter, since its resistance is extremely low and it is not the heating element. You will, however, have to anticipate the places that you will have to reach with it—under the dashboard, in and around the

PARTS LIST

5/16" - diameter compression fitting

Flashlight battery carbon, sharpened to a point at one end

Handle from file or other hand tool

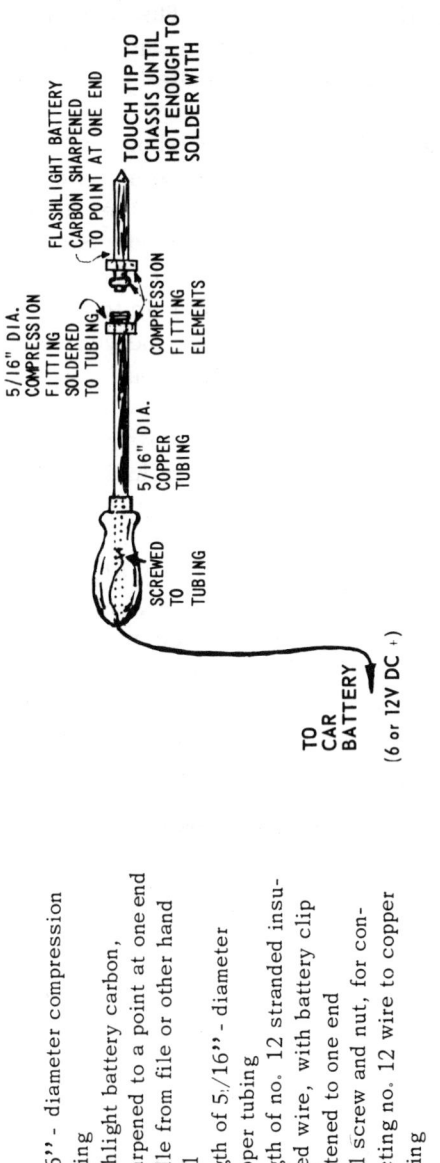

Length of 5/16" - diameter copper tubing

Length of no. 12 stranded insulated wire, with battery clip fastened to one end

Small screw and nut, for connecting no. 12 wire to copper tubing

---

TOUCH TIP TO CHASSIS UNTIL HOT ENOUGH TO SOLDER WITH

FLASHLIGHT BATTERY CARBON SHARPENED TO POINT AT ONE END

5/16" DIA. COMPRESSION FITTING SOLDERED TO TUBING

COMPRESSION FITTING ELEMENTS

5/16" DIA. COPPER TUBING

SCREWED TO TUBING

TO CAR BATTERY

(6 or 12V DC +)

engine, etc. A longer one is usually preferable. A piece of #12 stranded insulated wire with a battery clip is used as the power connection for the iron. This wire is connected to the tubing inside the handle with a small screw and nut. The tubing can be secured to the handle by jamming them together or using an adhesive.

The heating element is the carbon rod from an old flashlight battery, sharpened to a point at one end. To handle differences in diameter between the tubing and the carbon rod, a compression fitting is used. Solder the body of the fitting to the copper tubing. Seat the carbon rod as far as possible into the fitting and ferrule to insure a solid grip on the rod. Screw the nut into place and you're all set. The rod may be replaced occasionally, but you may need extra ferrules for this.

To use your iron, hook it to the "hot" terminal of a 6- or 12-volt battery. The other terminal of the battery goes to ground at the car chassis, so you can make contact anyplace on the metal surface of the car and current will flow. Since the carbon represents an extremely high resistance, it heats quite fast, glowing red in no time. Merely apply the solder and you've done what you had no way of doing before. The battery will be good for short soldering jobs, but not for any extended length of time. Accordingly, you have to be a little conservative; otherwise you may find yourself with a wonderfully soldered connection—and a dead battery.

52
Cheap Light Alarm

Far too common these days is an inoperable auto, lifeless only because its headlights were left on after the ignition was turned off. Rainy and early-morning-into-daylight driving have provided the setting for many such "slips." For the driver who prefers foresight and prevention rather than cure we present this ingeniously simple buzzer alarm that will at least make that driver aware that he has a potential power drain on his hands.

The alarm consists of three components plus some insulated connecting wire. Merely wire them in the specified order between the light switch and the **ignition** switch. If the polarity of the rectifier is reversed, you will have an alarm, but it will not be what you bargained for. The buzzer will sound if you do not turn off the lights and **ignition on** at the same time. Since the HF buzzer is a 6-volt type, Rl serves as a voltage dropping resistor in 12-volt systems.

When one, but not both, of the switches are on, the difference in potential generates a current flow. In one case this flow is blocked by the rectifier; in the other—the hazardous

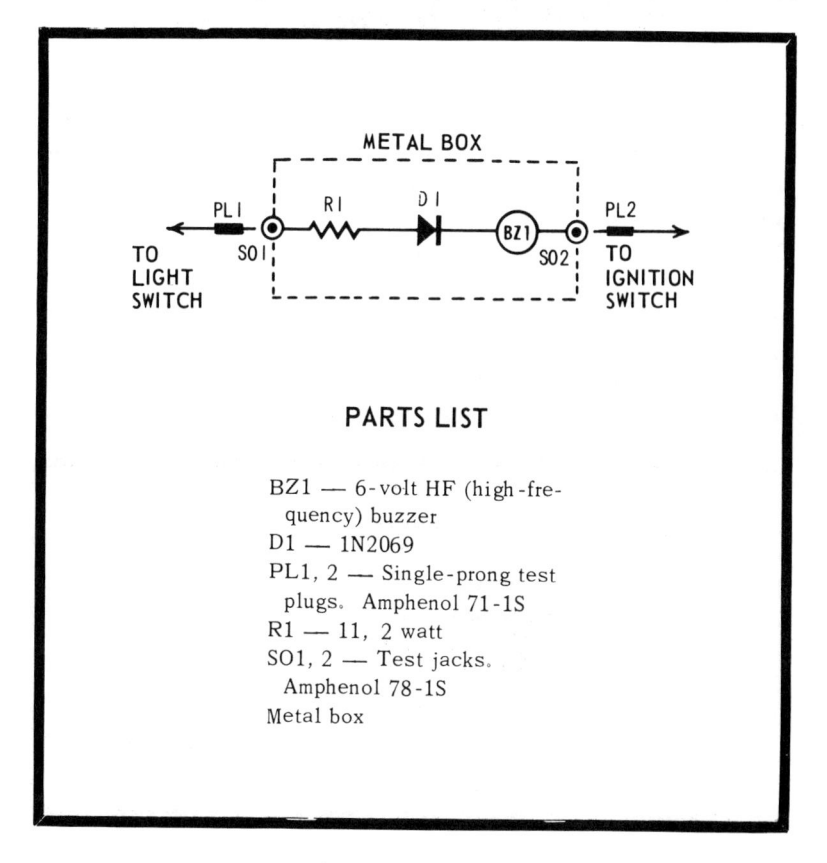

PARTS LIST

BZ1 — 6-volt HF (high-frequency) buzzer
D1 — 1N2069
PL1, 2 — Single-prong test plugs. Amphenol 71-1S
R1 — 11, 2 watt
SO1, 2 — Test jacks. Amphenol 78-1S
Metal box

165

one—current flows, operating the high-frequency buzzer. Perpetual freedom from the nagging fear of an inanimate auto is surely worth the price of this super-simple reminder alarm.

53

Automatic Garage Light

Groping and stumbling out of your car into a pitch black garage is no longer necessary with this automatic circuit which turns on your garage lights. The circuit responds to sound—whether it be car engine, horn, or even your voice. The 2-stage transistor amplifier controls a relay in the ceiling light circuit. The sound sensor is a speaker-turned-mike. An ordinary speaker is fine for this purpose. The circuit has its own step-down plug-in power supply.

R2 determines the level of sound required to operate your detector (the sensitivity control). An additional feature is

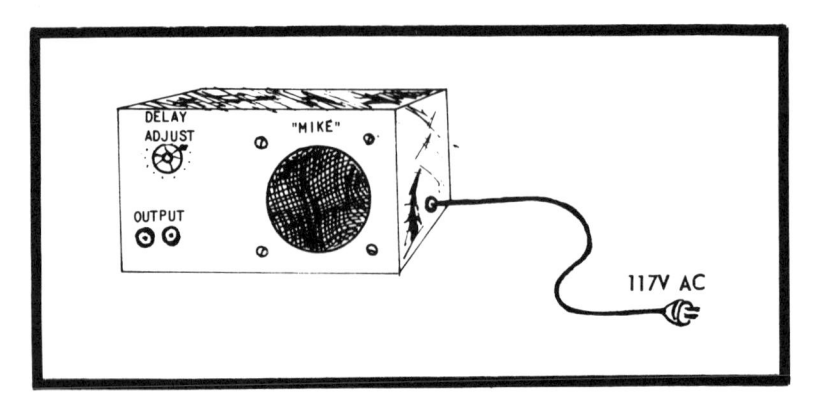

PARTS LIST

C1 — 50 mfd 15 WVDC electrolytic
C2 — 15 mfd, 15 WVDC electrolytic
C3, 4 — 1000 mfd, 15 WVDC electrolytic
D1, 2, 3 — 1N38B
K1 — 5K relay
PL1 — AC wall plug. Amphenol 61-F11
PL2, 3 — Single-prong test plug. Amphenol 71-1S
Q1, 2 — GE-2
R1 — 510K
R2 — 15K pot
R3 — 11K
R4 — 75K
SO1, 2 — Test jacks. Amphenol 78-1S
T1 — 117V AC primary; 12.6V DC center tapped secondary
"Junkbox" speaker

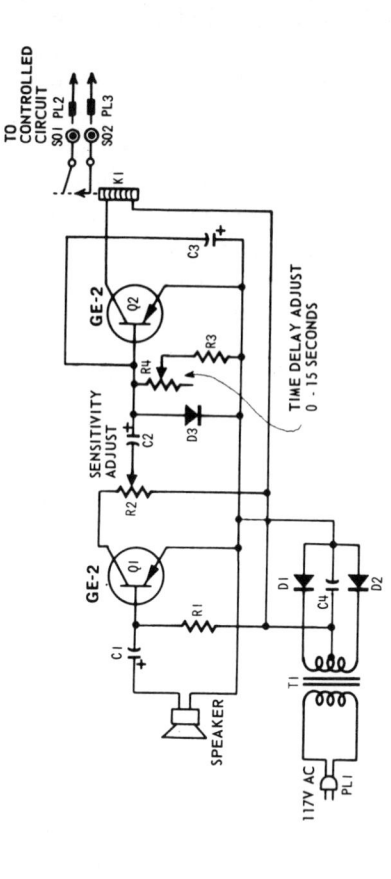

time-delay adjust R4, which allows you to choose when you would like the light to respond after the initial sound signal. Up to 15 seconds delay is possible. You can trigger the system as your car enters the garage and have the light go on as you're just leaving the car. Watch the polarities of the transistors, electrolytic capacitors, and diode. A small metal chassis is quite adequate for construction purposes.

Break into the light circuit with the two relay output leads. Although the automatic relay circuit is low voltage, the light is traditionally 117v AC and tricky to handle. Be extremely wary of any exposed AC wiring in your project. With the setup as it stands, a short sound means a short period of light. You might want to use a relay that remains closed until it is reset.

54
Economy Door/Hood/Trunk Alarm System

Nothing is more satisfying than finding a hidden and simple use for something that you already have. This alarm makes use of the car's already existing alarm device—the horn—plus the switches that monitor the opening and closing of your car doors and that trigger response of the dome light. Anyone entering your car unlawfully will create a ruckus that he will regret.

The only change in the circuit is a splice between the dome light circuit and the horn circuit, broken by a single switch for controlling the alarm. Whenever the alarm is on, closing any of the door or trunk switches—by opening the door—will trigger the horn relay as well as the dome light. No one could ask for anything simpler or more ingeniously effective.

```
                    HORN
                    RELAY                              TO "-" SIDE OF
                                         DOME          CAR BATTERY
                                         LIGHT
         CAR
         HORN

                         HORN                    DOOR         HOOD OR
                         BUTTON                  SWITCH       TRUNK
                                    DOOR                      SWITCH
                                    SWITCH
```

PARTS LIST

SW1 — SPST, Oak 200

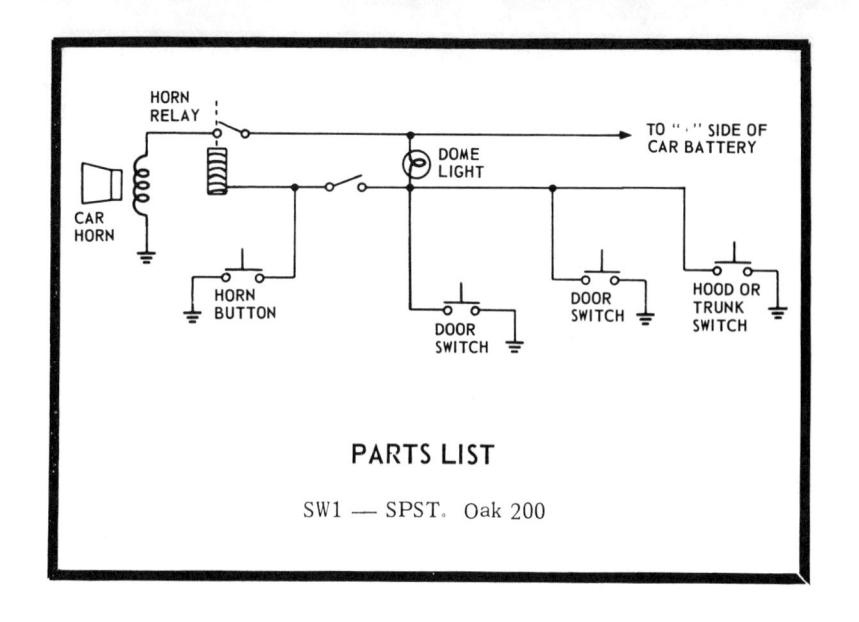

55

Mobile RF Power Meter & Dummy Load

An RF power meter requires a resistive load of known value and an ammeter. This project, of course, contains both. It also features two possible scales of magnitude and an unusually simple method of calibration. Any mobile transmitter with a rating of up to 100 watts is within range. After a transmitter is completely aligned, the finishing touch is often added with a power meter to assure maximum power output.

Ideal for the construction of your meter is a small, firmly grounded metal box. The input from your transmitter antenna is coaxially shielded, although an unshielded contact point should be easily accessible for alignment purposes. To

169

PARTS LIST

C1 — .0022 mfd
C2 — .02 mfd
D1, 2 — IN34A
I1 — Various household bulbs. See text
M1 — 0 - 200 DC microammeter
PL1 — AC wall plug. Amphenol 61-F11
R1 — 50 ohm, 40 watts (use combination of lesser values such as 20 1K, 2 watt resistors)
R2 — 360K
R3 — 500K, 5 watt pot
R4 — 36K
R5 — 110K
SO1 — Coax receptacle. Amphenol 83 -1R

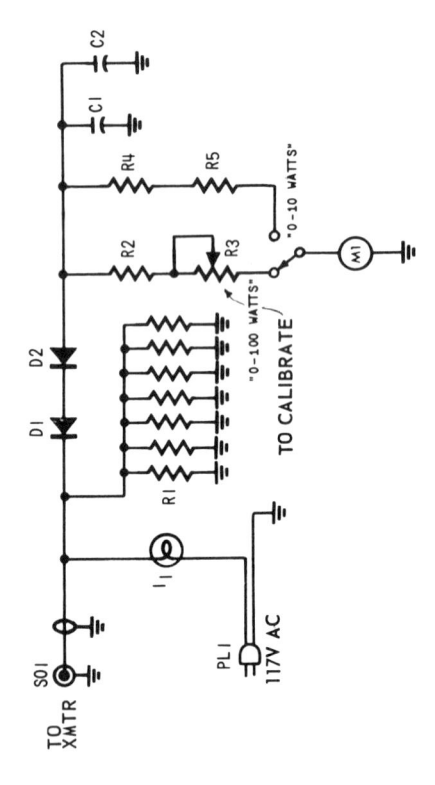

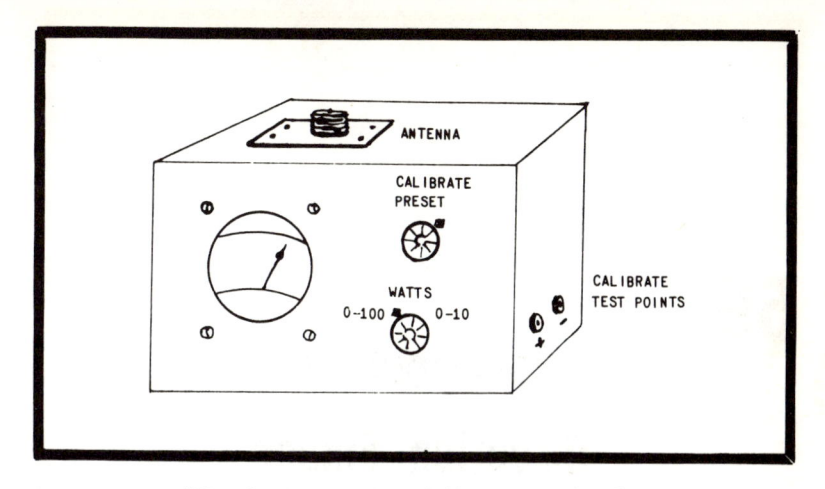

insure a quality instrument, follow standard construction procedure.

Although the value of your dummy load must be as specified, there are a great variety of methods of creating it. A small number of low resistance, high-wattage resistors may be replaced by a larger number of high-resistance, low-wattage resistors. Once you get your fixed resistance settled, the meter can be calibrated. Two rectifiers, which are far more efficient than one at higher frequencies, cut the signal in half. C1 and C2 smooth the pulsating DC to insure a more accurate "average" reading on the milliammeter.

Several different light bulbs of known wattage all under 100 watts, provide the key in calibrating. Hook one up as shown in the schematic, making use of the contact point at the antenna input. Here you will find that power cancels out power. If the bulb does not light at all when touched to the transmitter contact, note the reading on M1, and you have a calibration. Repeat for several wattages to find the correct range. R3 will give you a little more control over meter sensitivity and range. Because the power varies as the square of the current, the scale is not linear. Maximum power output is maintained with periodic meter checks against previous readings.

56
Foolproof CB Direction Finder

Talking to a CBer and driving towards him at the same time by means of a directional antenna, with the idea of surprising him by dropping in, will probably never cease to delight CB enthusiasts. This antenna, designed for 28 MHz, will not give you any bum leads, providing you get going in the right direction in the first place. It is highly directional in two directions, these being perpendicular to the plane of the loop.

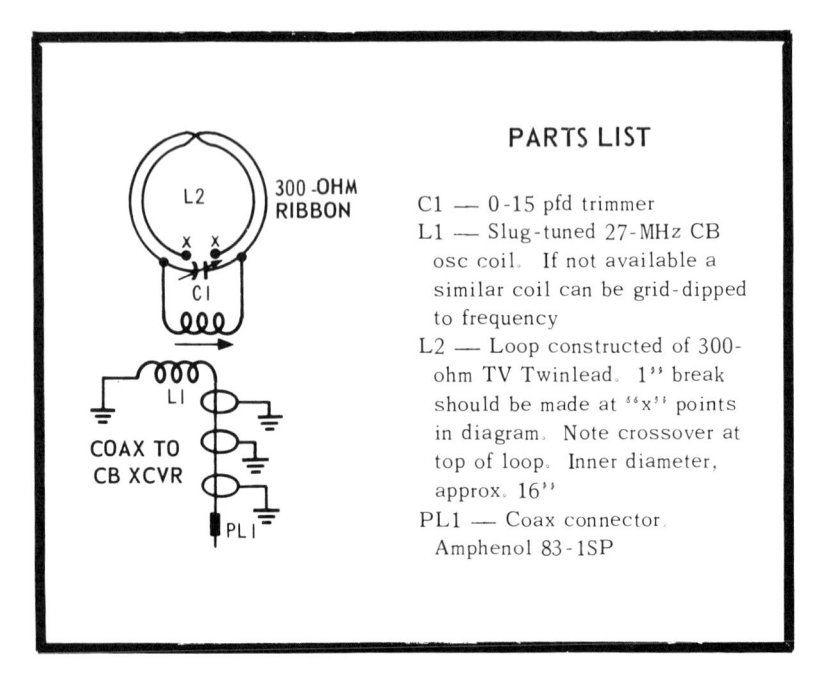

PARTS LIST

C1 — 0-15 pfd trimmer
L1 — Slug-tuned 27-MHz CB osc coil. If not available a similar coil can be grid-dipped to frequency
L2 — Loop constructed of 300-ohm TV Twinlead. 1" break should be made at "x" points in diagram. Note crossover at top of loop. Inner diameter, approx. 16"
PL1 — Coax connector. Amphenol 83-1SP

The heart of the project is a slug-tuned 27-MHz CB oscillator coil. This coil is connected, as shown in the pictorial, coaxially to the transceiver. If this specific coil is not available, a similar one may be grid-dipped to the correct frequency. A trimmer makes limited tuning possible.

TV twin-lead, the universal home-brew antenna material, comes to the rescue again. The main loop of this antenna is constructed of 300-ohm twin-lead. It should be twisted once at the top and broken where shown by the Xs in the diagram. The inner diameter of the loop is about 16".

The antenna may be permanently fixed with a wooden or metallic pole to the outside of your car, or it can be mounted loosely so it can be turned. When the antenna is fixed, heading the car in different directions will indicate directional signal strength.

Because of the duplicity of your direction finder, an S-meter for your set may help in estimating signal strength. If you don't already have one, we invite your attention to the one described in Project 58.

57

Vibrator Rejuvenator

Essential to the operation of a vibrator is the condition of its iron-core coils. An iron core is used because of its resistance to permanent magnetism and because of its extremely good permeability to magnetic flux as opposed to that of air. In a vibrator the current through these coils flows in only one

PARTS LIST

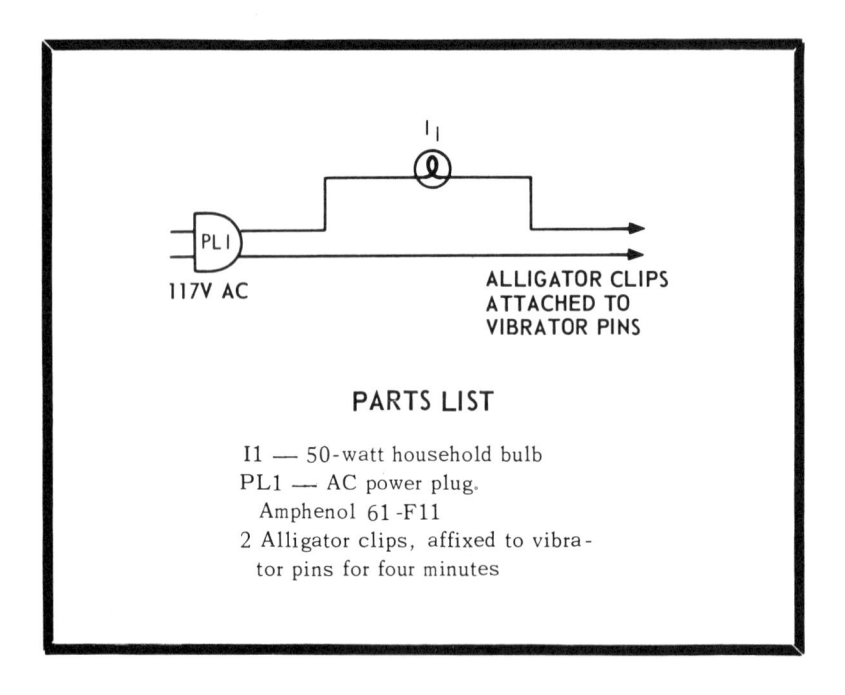

I1 — 50-watt household bulb
PL1 — AC power plug.
  Amphenol 61-F11
2 Alligator clips, affixed to vibrator pins for four minutes

direction. The core, subjected to a magnetic field which is always of the same polarity, tends to loose some of its permeability as a result of a phenomenon called "saturation." Single-ended amplifier stages often do not use transformer coupling for this reason. A vibrator tends to "run-down" when its cores become saturated. If you have this problem with your vibrator, here is what you can do about it.

The rejuvenator consists only of AC line cord with a plug and alligator clips and a 50-watt light bulb. Put the bulb in one of the lines, hook the clips to the vibrator pins, plug it in for about four minutes, and the vibrator will be as good as new. The AC voltage applied to the coils tends to break the saturation by constantly reversing the magnetic field passing through the cores.

The vibrator will be in no way damaged by this treatment, since there are no stabilizing or rectifying elements in the input stage. It is apt to heat up a little, but the short treatment takes care of that.

58

"S" Meter For Mobile Receivers

An accurate indication of incoming signal strength is always helpful in getting right "on" a station and staying there. It tells you how well the fellow at the other end is getting through. If he informs you that his set is working perfectly, but your meter says not, then your "S" meter may be indicating a flaw in your own set. It is just another useful device that the mobile enthusiast should have at his disposal. This circuit is a very fundamental "S" meter.

Tie into your receiver at the DC plate lead of the AGC-controlled IF stage. The circuit contains a milliammeter arranged in a bridge so that readings increase with AVC voltage and signal strength. The meter and potentiometer R1 can be conveniently mounted in a small case and placed on the dashboard, within easy sight of the operator. Be sure to

175

PARTS LIST

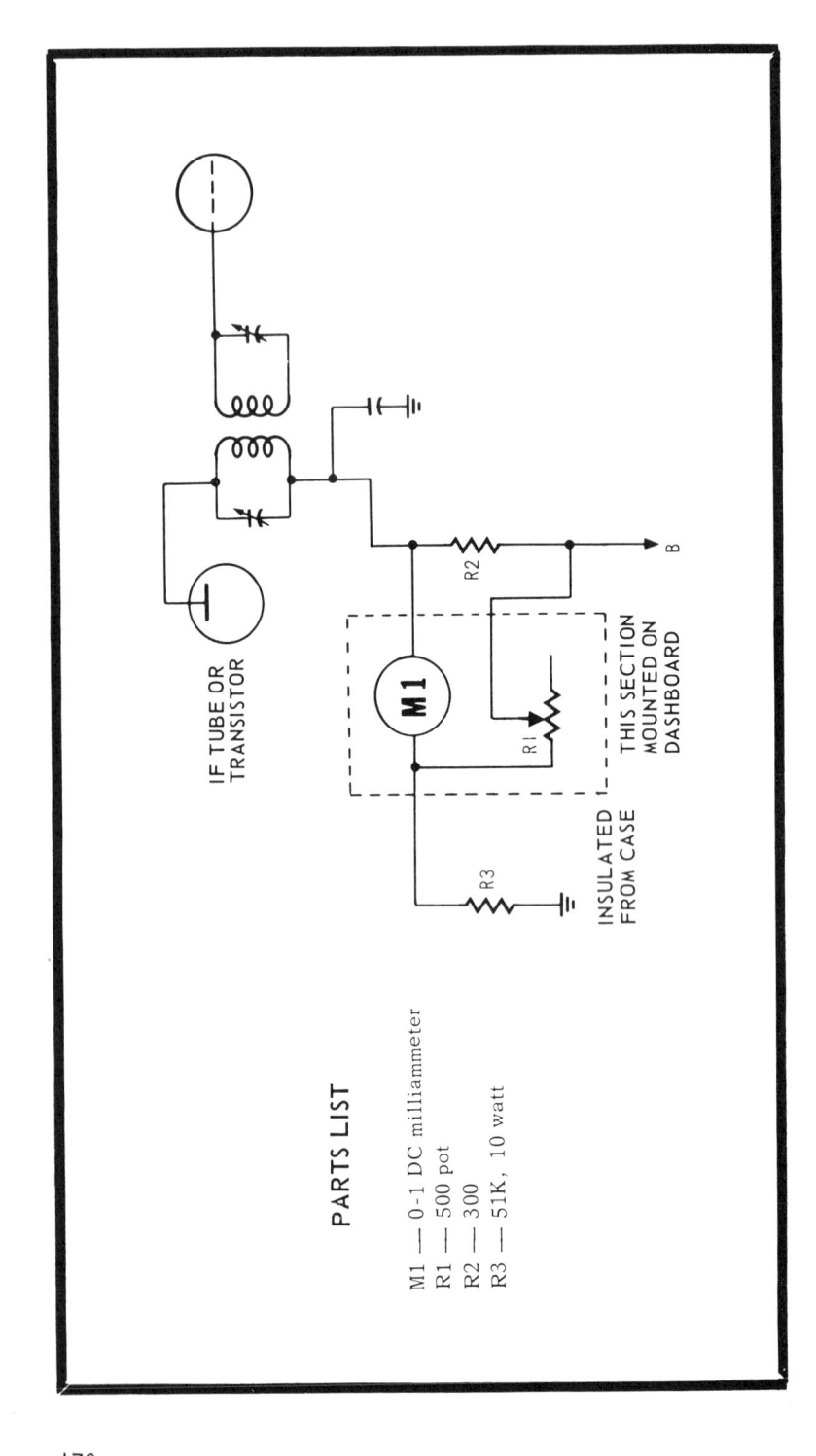

M1 — 0-1 DC milliammeter
R1 — 500 pot
R2 — 300
R3 — 51K, 10 watt

insulate R1 from the case. The other components are connected inside the receiver.

To adjust your meter, pull the tube out of its socket or open the cathode circuit so that no plate current can flow. Adjust R1 until the scale reading reaches maximum—your highest meter reading. (The value of the bridge resistor depends on the internal resistance of the meter.) Then stick the tube back in and try to find a happy medium with R1. The scale does not have to reach zero for it to be a good tuning indicator. If you adjust your meter down to zero, the response will not be as good. For local work, however, it will not make a great deal of difference. The meter reads on an almost linear decibel scale.

59
Delay-Action Foilproof Burglar Alarm

This is not just another burglar alarm. It makes use of the conventional horn and the already existing door switches, but it has an extra feature that tremendously increases its effectiveness. It does not go off when the person initially enters the car unlawfully, but only after he is thoroughly entrenched in his ill-deed. Picture the burglar: If he intends to take something or perhaps steal the car itself, he is already quite nervous and on edge. As his act progresses to its final stage, he probably is a real bundle of nerves. If at that time the horn starts blowing, he will probably jump right out of his skin.

PARTS LIST

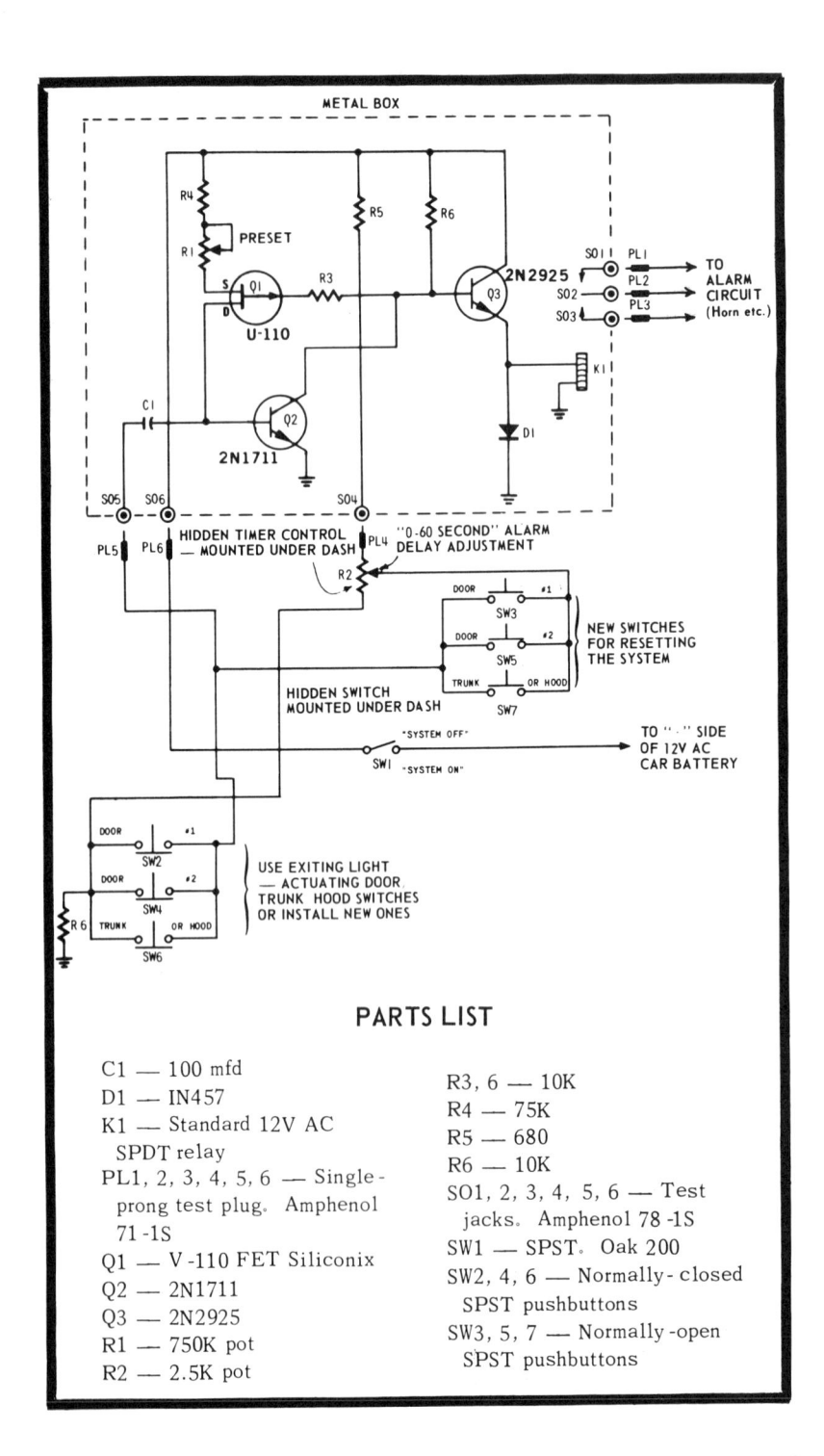

C1 — 100 mfd
D1 — IN457
K1 — Standard 12V AC SPDT relay
PL1, 2, 3, 4, 5, 6 — Single-prong test plug. Amphenol 71-1S
Q1 — V-110 FET Siliconix
Q2 — 2N1711
Q3 — 2N2925
R1 — 750K pot
R2 — 2.5K pot
R3, 6 — 10K
R4 — 75K
R5 — 680
R6 — 10K
SO1, 2, 3, 4, 5, 6 — Test jacks. Amphenol 78-1S
SW1 — SPST. Oak 200
SW2, 4, 6 — Normally-closed SPST pushbuttons
SW3, 5, 7 — Normally-open SPST pushbuttons

The alarm-tripping switches are the normally-closed type and already in the dome light circuit, also. Three new switches are added for resetting the system. These switches are not absolutely necessary. A 3-transistor configuration provides the delay before the relay is activated. The main components are placed in a small box behind the dash, with switch SW1 (which turns the alarm on) and the other adjustment controls in a concealed location, accessible only to the person who knows where they are.

The circuit is powered by the car battery, and the relay is connected into the horn circuit, past the horn button, of course. A 0- to 60-second delay may be obtained by adjusting R2. R1 must be carefully preset for the relay to operate.

60

Generator "Hash" Eliminator

A considerable amount of undesirable noise and static is produced by the brush contacts in the generator of your car. Some systems are carefully shielded to prevent this "hash" from getting out. If, however, you have been noticing a lot of "garbage" on your car radio lately, it may be hash from the generator. The problem can be solved quite simply by introducing two components into your electrical system.

A parallel resonant circuit, consisting of a trimmer capacitor and a home-brew coil, is all you need. The directions for constructing the coil are given in the parts list. In parallel,

TO ARMATURE LEAD   TO ARMATURE TERMINAL

L1

TO ARMATURE LEAD   C1   TO ARMATURE TERMINAL

PARTS LIST

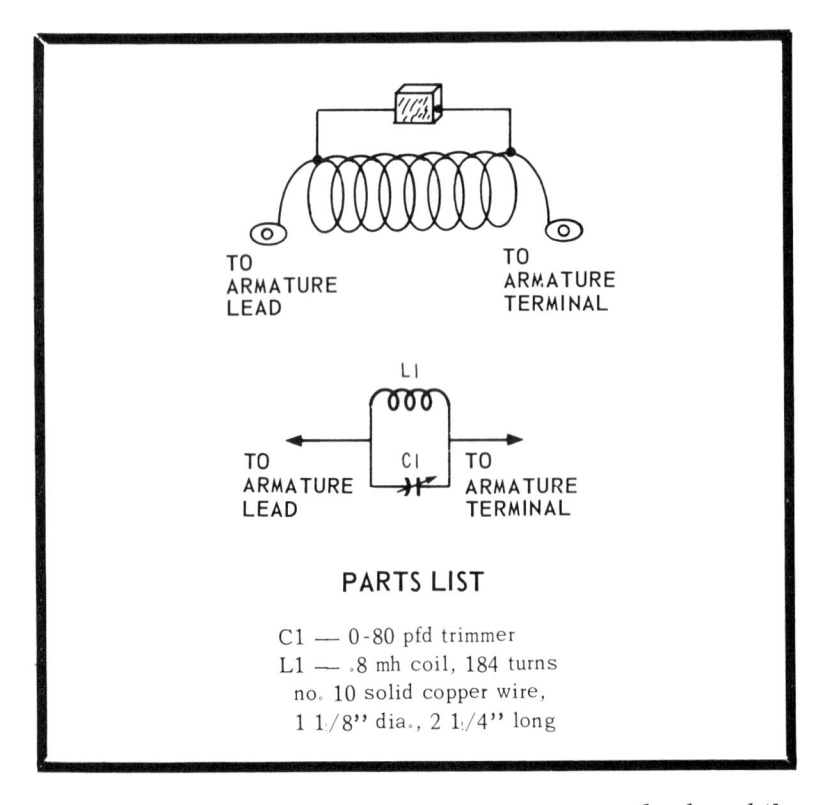

C1 — 0-80 pfd trimmer
L1 — .8 mh coil, 184 turns
no. 10 solid copper wire,
1 1/8" dia., 2 1/4" long

the two are inserted between the armature lead and the armature terminal. The hash is essentially "turned out," Adjust C1 for optimum results.

61

Handy Timing Light

A strobe light is as essential to the maintenance of your automobile as are the tachometer and the dwell meter. With

CONNECT TO BAR CONNECTING CELL 3 WITH CELL 4 IN 12-VOLT BATTERY OR INSTALL 15-OHM 100-WATT POT AND ADJUST FOR 6V DC AT THIS POINT

PARTS LIST

C1 — .005 mfd 1500-volt rating
C2 — 10 mfd, 750 WVDC electrolytic
C3 — 2 mfd, 750 WVDC electrolytic
D1 — 1000-volt selenium. Sarkes-Tarzian HV Type I
PL1 — Single-prong test plug. Amphenol 71-1S
R1 — 270, 2 watts
R2 — 27K, 2 watts
SO1 — Test jack. Amphenol 78-1S
T1 — 500V AC, CT vibrator type. Thordarson 22R26
VB1 — 6-volt nonsynchronous vibrator. Mallory 509P
X1 — Strobe tube, xeon gas type. Amglo U-35

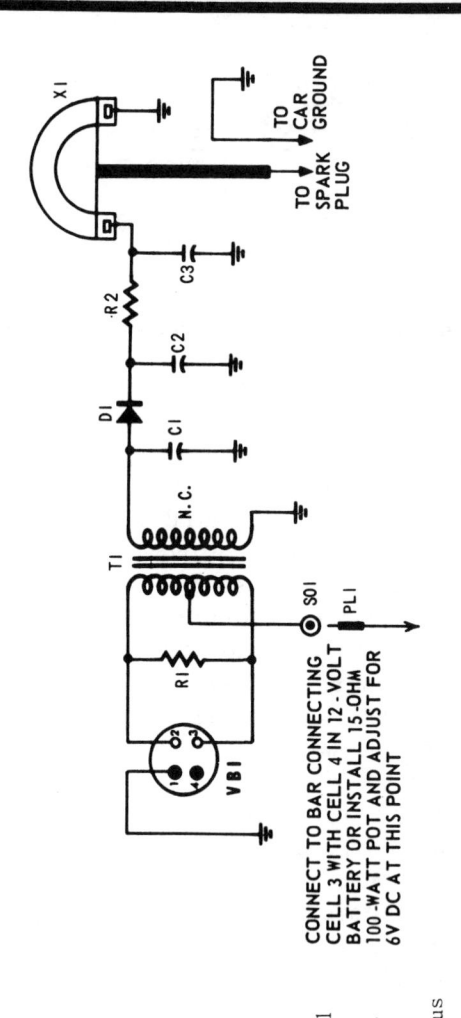

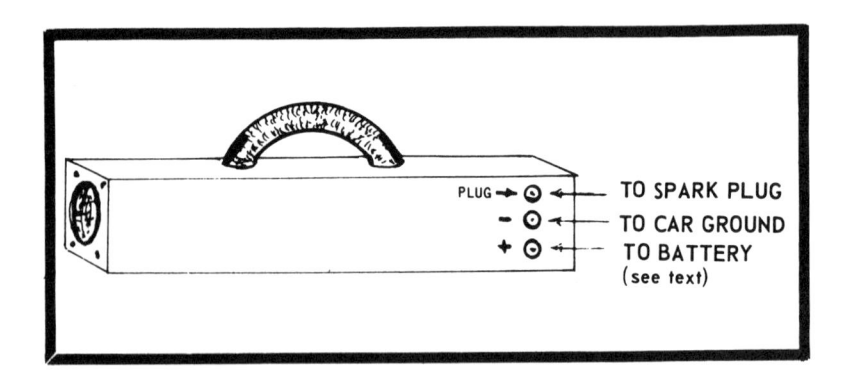

```
                                    PLUG → ⊙ ←       TO SPARK PLUG
                                         − ⊙ ←       TO CAR GROUND
                                         + ⊙ ←       TO BATTERY
                                                      (see text)
```

it, engine timing can be set to perfection. This factor plays a great part in determining your engine's operating efficiency. You get a visual indication of the firing of a spark plug in relation to the overall engine cycle.

Providing the necessary potential across the strobe tube is a nonsynchronous vibrator. Although each voltage pulse is not uniform, the frequency is so great compared to that of the spark plug that it makes little difference. The vibrator is a 6-volt type and is coupled into the circuit with a special vibrator transformer. To get the required 6 volts, tap the bar connecting cell #3 with cell #4 on a 12-volt battery, or use a 15-ohm, 100-watt potentiometer to drop the battery voltage to 6 volts.

The entire circuit, except for the light, may be closed in an insulating metal box, using leads and battery clips for the external connections. Watch the polarity of the diode and the electrolytic capacitors.

To use the timing light, connect it to the battery and spark plug. Optimum efficiency is obtained by retarding the spark, which is done by loosening the distributor cap and rotating the distributor in the direction of operating rotation. About 40 RPM short of maximum retardation will give best efficiency. Check this with a tachometer, if possible.

62

Accurate Miles-Per-Gallon Meter

Here is a two-in-one deal that will interest you. One circuit is a miles-per-gallon meter, the other a tachometer. Together they function as sort of a universal tune-up kit. In either case, a Minibox is perfect for construction of the circuit. The meter, selector switch SW1, and R8 are mounted in one side of the box. Leads ending in alligator clips run directly into the box for the battery and coil connections. Two terminals for R9 are needed also.

R9 is the key in the mpg circuit. It is mounted on an L-bracket with the shaft free to turn below the gas pedal of the car. A homemade coupling arm made of scrap metal is linked through a shaft coupler to the potentiometer, as shown in the pictorial. Providing tension is a spring: as the pedal is pushed down, the shaft of the resistor turns and, conse-

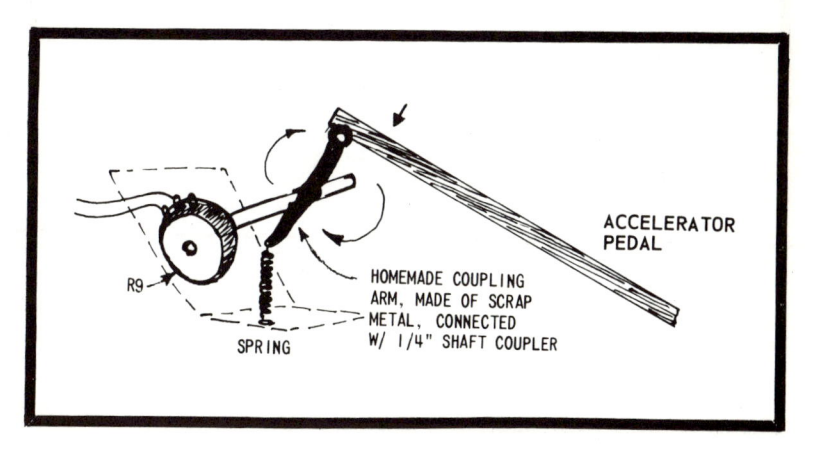

183

PARTS LIST

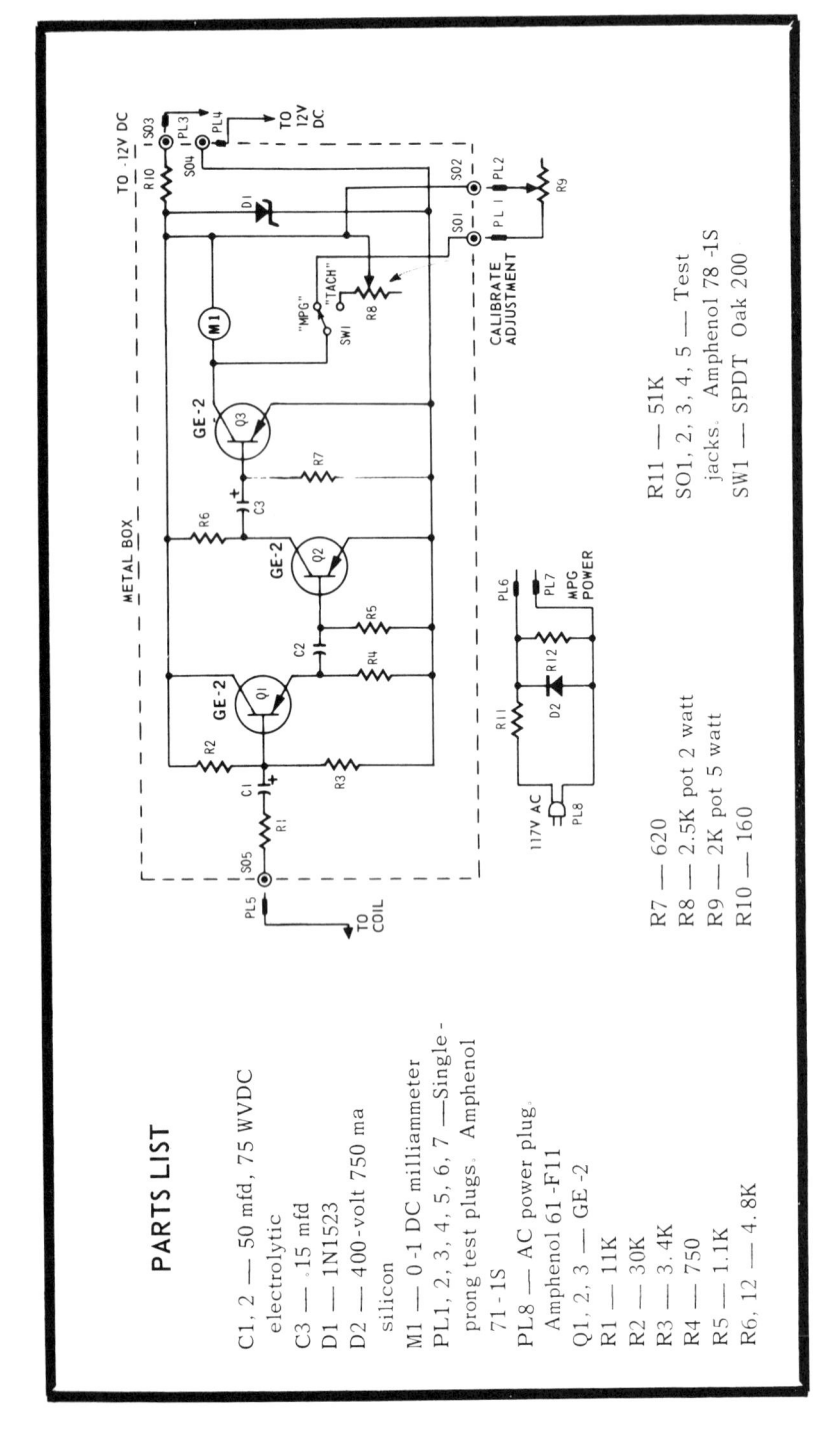

C1, 2 — 50 mfd, 75 WVDC electrolytic
C3 — .15 mfd
D1 — 1N1523
D2 — 400-volt 750 ma silicon
M1 — 0-1 DC milliammeter
PL1, 2, 3, 4, 5, 6, 7 — Single-prong test plugs. Amphenol 71-1S
PL8 — AC power plug. Amphenol 61-F11
Q1, 2, 3 — GE-2
R1 — 11K
R2 — 30K
R3 — 3.4K
R4 — 750
R5 — 1.1K
R6, 12 — 4.8K
R7 — 620
R8 — 2.5K pot 2 watt
R9 — 2K pot 5 watt
R10 — 160
R11 — 51K
SO1, 2, 3, 4, 5 — Test jacks. Amphenol 78-1S
SW1 — SPDT Oak 200

quently, the reading on M1 changes. The assembly can be left there but must be disconnected when not in use. 117v AC is connected at points 1 and 2. There are no 12-volt connections.

To use the mpg meter, hook up the R9 assembly and the power. Set R9 to calibrate the meter in some arbitrary way. The meter is primarily useful for periodic checks of engine efficiency. Record the reading at several different RPM's immediately after a tune-up. Check it a few weeks later and note any change at the same RPM's.

To use the tachometer, turn the selector to "tach," make the two battery connections, as shown, and the connection to the distributor points terminal (coil). Align the meter with R8. With your tachometer you can set the carburetor, check engine timing, and check the transmission, and it even has several other useful functions. For a more detailed discussion of its uses and operation, see the project entitled "Subminiature Tachometer."

63

Double-Purpose Siren

Nothing is more frightening to a burglar than a siren. This one is wired so that if any one of the doors that control the ceiling light is opened, the siren will let out a yowl that may wake the dead. The siren also serves another purpose, that being to draw enough current to activate a relay connected in the circuit. If the relay is connected to the car horn, no burglar is going to linger for very long after he opens the door. Even if the door is closed right away, it is long enough to let everybody within blocks know that he is there.

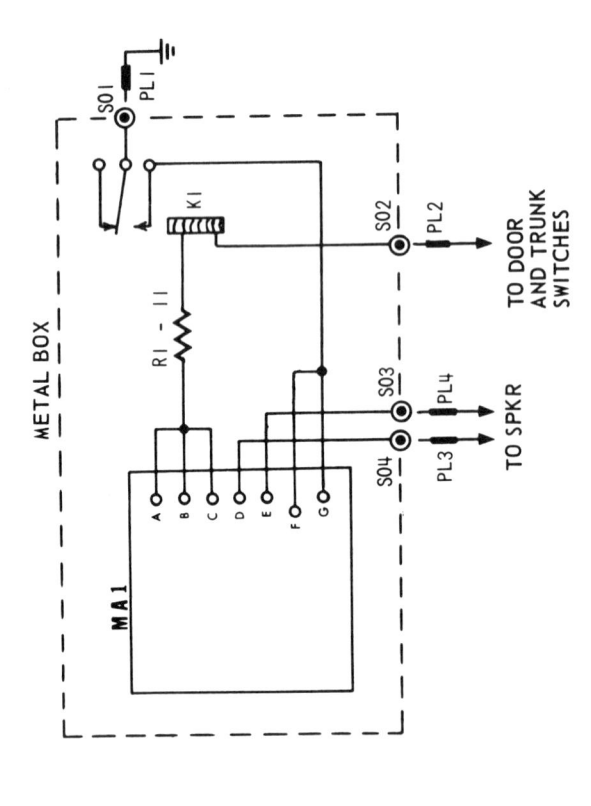

PARTS LIST

K1 — SPDT 12-volt relay. Potter & Brumfield. RS5D 12V DC

MA1 — Sirenmodule. Eicocraft EC-100

PL1, 2, 3, 4 — Single-prong test plug. Amphenol 71-1S

R1 — 11, 10 watt

SO1, 2, 3, 4 — Test jacks. Amphenol 78-1S

A siren module is connected to the door and trunk switches. You may want to put the speaker up by the grill to make the siren easier to hear. The horn type is good for this particular application. The circuit operates on 12 volts. Make whatever connections you like to the relay. Mount the siren module and resistor behind the dash. If you happen to be a volunteer fireman, you may have additional uses for the siren (if lawful in your community).

64

Turn Signal Beeper

Too many drivers stake their cars and lives on the other fellow's turn signal. They drive along depending on the other car with his turn signal on to do just that—turn. The result? Head-ons, sideswipes, destruction, death. Sometimes a person is just not aware that his signal is on. Perhaps it was turned on to switch lanes, or make a partial turn. The signal indicators on many cars—usually a small flashing light, and an obscure clicking sound—are just not designed to really command the attention of the driver. For drivers, especially those whose hearing may not be what it used to be, here is a turn signal beeper that at least will make you aware that your signal is on.

The circuit operates on 12 volts and can be mounted in a small metal box and attached beneath the dash. A small speaker attached to the side of the box will be quite adequate

PARTS LIST

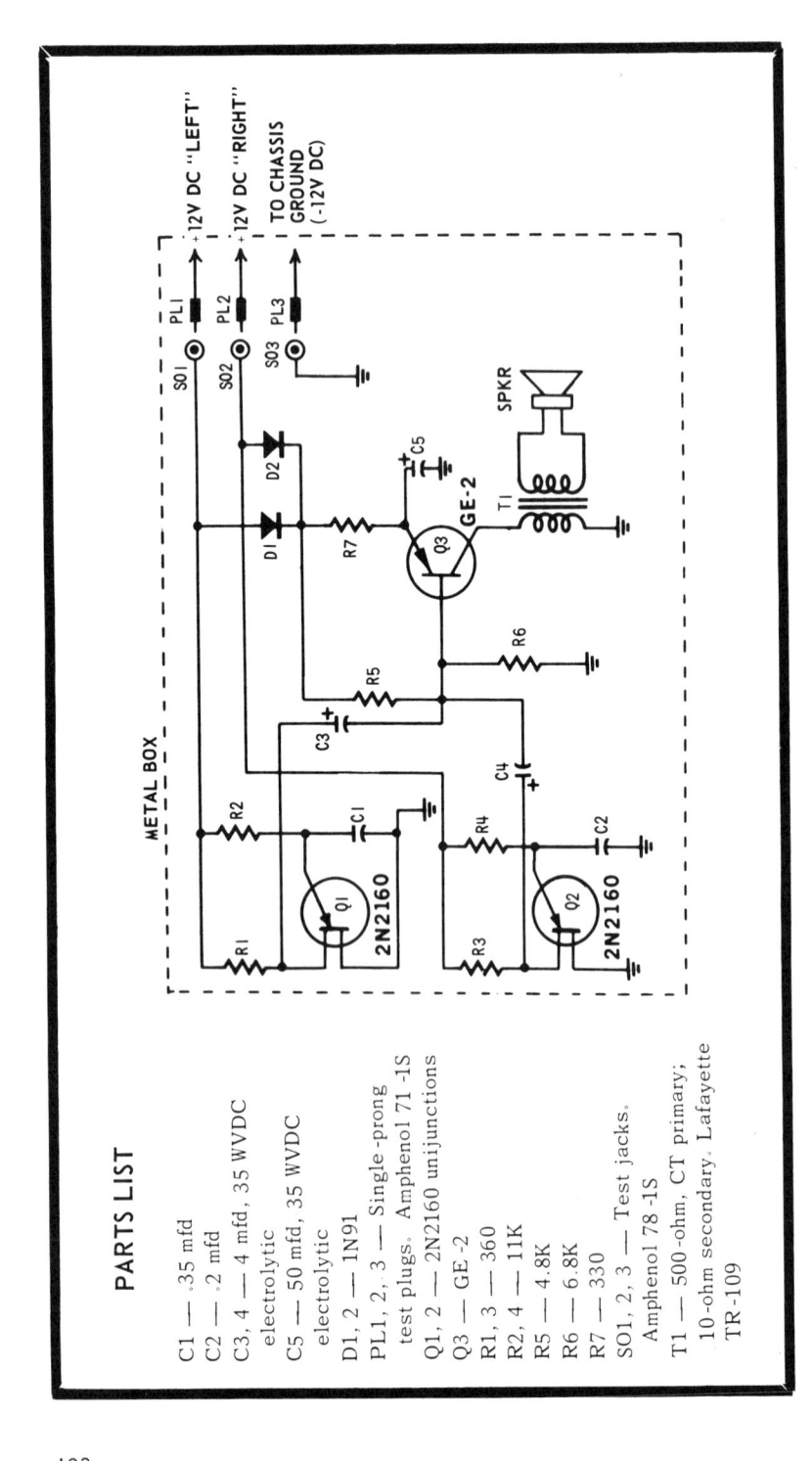

C1 — .35 mfd
C2 — .2 mfd
C3, 4 — 4 mfd, 35 WVDC electrolytic
C5 — 50 mfd, 35 WVDC electrolytic
D1, 2 — 1N91
PL1, 2, 3 — Single-prong test plugs. Amphenol 71-1S
Q1, 2 — 2N2160 unijunctions
Q3 — GE-2
R1, 3 — 360
R2, 4 — 11K
R5 — 4.8K
R6 — 6.8K
R7 — 330
SO1, 2, 3 — Test jacks. Amphenol 78-1S
T1 — 500-ohm, CT primary; 10-ohm secondary. Lafayette TR-109

to broadcast the "beep," unless you prefer a 12" woofer to really "sock" the signal to you.

The input leads for the beeper are the 12-volt output of the signal circuits—one left, one right. The circuit is actually two oscillators kicked into oscillation by the applied voltage when either signal is turned on.

Index

A

AC inverter, 140
Anti-theft device, 136
Attic fan control, 99
Auto ice alarm, 143
Auto light "minder," 155
Auto lock alarm system, 168
Automatic garage light, 166
Auto PA system, 157
Auto radio tape player "retriever," 131
Auto safety flasher, 128
Auto sleep alarm, 119
Auto soldering gun, 162
Auto theft "preventer," 136

B

Battery charger, 159
Burglar alarm, 64, 177

C

CB direction finder, 172
Closet light control, 56
Coin-operated switch, 105
Continuity tester, 72

D

Delay-action burglar alarm, 177
Detector, rain, 67
Direction finder, 121, 153, 172
Door/hood/trunk lock alarm, 168
Door lock, 90
Double-purpose siren, 185
Drill switch, 75
Dummy load, 169
Dwell meter, 133, 151

E

Electric candle, 80
Electric door lock, 90

F

Fire alarm, 37
Flasher, auto, 128

G

Garage light control, 31, 166
Generator "hash" eliminator, 179

H

"Hash" eliminator, 179
Headlight alarm, 117

I

Ice alarm, 143
Idle speed calibrator, 123
Inverter, 140
Interference killer, 135

L

Lamp control, 31, 41, 53, 87, 108
Lap counter, 61
Lawn sprinkler, 44
Light alarm, 164
Light control, 31, 56
Light "minder," 155
Liquid level control, 48
Lock alarm, 168
Low-voltage switch, 56

M

Mat switch, 102
Miles-per-gallon meter, 183
Mobile power supply, 145
Mobile voice control, 137
Motor controller, 77

P

PA system, 157
Power meter, RF, 169
Power supply, 145
Power supply overload protector, 85
Press-to-talk switch, 96

R

Radio & tape player "retriever," 131
Rain detector, 67
Recorder switch, 96
Regulator interference killer, 135
Remote switch, 53, 83
RF power meter, 169

S

Safety flasher, 128
Shooting gallery, 69
Siren, 185
Sleep alarm, 119
Slot car lamp counter, 61
"S" meter, 175
Smoke alarm, 110
Soldering gun, 162
Soldering iron control, 58
Splash alarm, 93
Sprinkler control, 44
Swimming pool alarm, 93

T

Tachometer, 115, 123, 148
Tape recorder switch, 96
Thermostat control, 99
Timing light, 180
Transmitter direction finder, 121
Turn signal beeper, 187
TV control, 35, 53, 105

V

Vibrator rejuvenator, 173
Vibrator "substitutor," 130
Voice control, 137
VOM-to-Dwell meter converter, 133

W

Wake-up alarm, 119